Communications in Computer and Information Science   1739

More information about this series at https://link.springer.com/bookseries/7899

Nicholas Olenev · Yuri Evtushenko ·
Milojica Jaćimović · Michael Khachay ·
Vlasta Malkova · Igor Pospelov (Eds.)

Advances in Optimization and Applications

13th International Conference, OPTIMA 2022
Petrovac, Montenegro, September 26–30, 2022
Revised Selected Papers

Editors
Nicholas Olenev (iD)
FRC CSC RAS
Moscow, Russia

Yuri Evtushenko (iD)
FRC CSC RAS
Moscow, Russia

Milojica Jaćimović (iD)
University of Montenegro
Podgorica, Montenegro

Michael Khachay (iD)
Krasovsky Institute of Mathematics
and Mechanics
Ekaterinburg, Russia

Vlasta Malkova (iD)
FRC CSC RAS
Moscow, Russia

Igor Pospelov (iD)
FRC CSC RAS
Moscow, Russia

ISSN 1865-0929   ISSN 1865-0937 (electronic)
Communications in Computer and Information Science
ISBN 978-3-031-22989-3   ISBN 978-3-031-22990-9 (eBook)
https://doi.org/10.1007/978-3-031-22990-9

This Springer imprint is published by the registered company Springer Nature Switzerland AG
The registered company address is: Gewerbestrasse 11, 6330 Cham, Switzerland

Preface

This volume contains the second part of the refereed proceedings of the XIII International Conference on Optimization and Applications (OPTIMA 2022)[1].

Organized annually since 2009, the conference has attracted a significant number of researchers, academics, and specialists in many fields of optimization, operations research, optimal control, game theory, and their numerous applications in practical problems of data analysis and software development.

The broad scope of OPTIMA has made it an event where researchers involved in different domains of optimization theory and numerical methods, investigating continuous and discrete extremal problems, designing heuristics and algorithms with theoretical bounds, developing optimization software, and applying optimization techniques to highly relevant practical problems can meet together and discuss their approaches and results. We strongly believe that this facilitates collaboration between researchers working in optimization theory, methods, and applications, and those employing them to resolve valuable practical problems.

The conference was held during September 26–30, 2022, in Petrovac, Montenegro, in the picturesque Budvanian riviera on the azure Adriatic coast. For those who were not able to come to Montenegro this year, an online session was organized. The main organizers of the conference were the Montenegrin Academy of Sciences and Arts, Montenegro, the Dorodnicyn Computing Centre, FRC CSC RAS, Russia, and the University of Évora, Portugal. This year, the key topics of OPTIMA were grouped into seven tracks:

1. Mathematical programming
2. Global optimization
3. Discrete and combinatorial optimization
4. Optimal control
5. Optimization and data analysis
6. Game theory and mathematical economics
7. Applications

The Program Committee (PC) and invited reviewers included more than one hundred well-known experts in continuous and discrete optimization, optimal control and game theory, data analysis, mathematical economy, and related areas from leading institutions of 25 countries: Argentina, Australia, Austria, Belgium, China, Finland, France, Germany, Greece, India, Israel, Italy, Lithuania, Kazakhstan, Mexico, Montenegro, the Netherlands, Poland, Portugal, Russia, Serbia, Sweden, Taiwan, the UK, and the USA. This year we received 70 submissions, mostly from Russia but also from Azerbaijan, Kazakhstan, Latvia, Montenegro, Poland, Portugal, and the USA. Each submission was reviewed in a single-blind blind manner by at least three PC members or invited reviewers, experts in their fields, to supply detailed and helpful comments. Out of 43 qualified

---

[1] http://agora.guru.ru/display.php?conf=OPTIMA-2022.

submissions, the Program Committee decided to accept 17 papers to the first volume of the proceedings for publication in LNCS volume 13781. Thus the acceptance rate for the volume was about 40%.

In addition, after a short presentation of the candidate submissions, discussion at the conference, and subsequent revision, the Program Committee proposed 13 out of the remaining 26 papers to be included in this, second, volume of proceedings.

The conference featured two invited lecturers, plus several plenary and keynote talks. The invited lectures were as follows:

- Panos M. Pardalos, University of Florida, USA, "Computational Approaches for Solving Systems of Nonlinear Equations"
- Alexey Tret'yakov, Siedlce University of Natural Sciences and Humanities, Poland, "Degenerate Equality Constrained Optimization Problems and P-Regularity Theory"

We would like to thank all the authors for submitting their papers and the members of the PC for their efforts in providing exhaustive reviews. We would also like to express special gratitude to all the invited lecturers and plenary speakers.

October 2022

Nicholas Olenev
Yuri Evtushenko
Milojica Jaćimović
Michael Khachay
Vlasta Malkova
Igor Pospelov

Organization

Program Committee Chairs

Milojica Jaćimović	Montenegrin Academy of Sciences and Arts, Montenegro
Yuri G. Evtushenko	Dorodnicyn Computing Centre, FRC CSC RAS, Russia
Igor G. Pospelov	Dorodnicyn Computing Centre, FRC CSC RAS, Russia
Michael Yu. Khachay	Krasovsky Institute of Mathematics and Mechanics, Russia
Vlasta U. Malkova	Dorodnicyn Computing Centre, FRC CSC RAS, Russia
Nicholas N. Olenev	CEDIMES-Russie and Dorodnicyn Computing Centre, FRC CSC RAS, Russia

Program Committee

Majid Abbasov	St. Petersburg State University, Russia
Samir Adly	University of Limoges, France
Kamil Aida-Zade	Institute of Control Systems of ANAS, Azerbaijan
Alla Albu	Dorodnicyn Computing Centre, FRC CSC RAS, Russia
Alexander P. Afanasiev	Institute for Information Transmission Problems, RAS, Russia
Yedilkhan Amirgaliyev	Suleyman Demirel University, Kazakhstan
Anatoly S. Antipin	Dorodnicyn Computing Centre, FRC CSC RAS, Russia
Adil Bagirov	Federation University, Australia
Artem Baklanov	International Institute for Applied Systems Analysis, Austria
Evripidis Bampis	LIP6, Sorbonne Université, France
Olga Battaïa	ISAE-SUPAERO, France
Armen Beklaryan	National Research University Higher School of Economics, Russia
Vladimir Beresnev	Sobolev Institute of Mathematics, Russia
Anton Bondarev	Xi'an Jiaotong-Liverpool University, China
Sergiy Butenko	Texas A&M University, USA
Vladimir Bushenkov	University of Évora, Portugal

Igor A. Bykadorov	Sobolev Institute of Mathematics, Russia
Alexey Chernov	Moscow Institute of Physics and Technology, Russia
Duc-Cuong Dang	INESC TEC, Portugal
Tatjana Davidovic	Mathematical Institute of Serbian Academy of Sciences and Arts, Serbia
Stephan Dempe	TU Bergakademie Freiberg, Germany
Askhat Diveev	FRC CSC RAS and RUDN University, Russia
Alexandre Dolgui	IMT Atlantique, LS2N, CNRS, France
Olga Druzhinina	FRC CSC RAS, Russia
Anton Eremeev	Omsk Division of Sobolev Institute of Mathematics, SB RAS, Russia
Adil Erzin	Novosibirsk State University, Russia
Francisco Facchinei	Sapienza University of Rome, Italy
Vladimir Garanzha	Dorodnicyn Computing Centre, FRC CSC RAS, Russia
Alexander V. Gasnikov	National Research University Higher School of Economics, Russia
Manlio Gaudioso	Universita della Calabria, Italy
Alexander Yu. Gornov	Institute of System Dynamics and Control Theory, SB RAS, Russia
Edward Kh. Gimadi	Sobolev Institute of Mathematics, SB RAS, Russia
Andrei Gorchakov	Dorodnicyn Computing Centre, FRC CSC RAS, Russia
Alexander Grigoriev	Maastricht University, The Netherlands
Mikhail Gusev N.N.	Krasovskii Institute of Mathematics and Mechanics, Russia
Vladimir Jaćimović	University of Montenegro, Montenegro
Vyacheslav Kalashnikov	ITESM, Monterrey, Mexico
Maksat Kalimoldayev	Institute of Information and Computational Technologies, Kazakhstan
Valeriy Kalyagin	Higher School of Economics, Russia
Igor E. Kaporin	Dorodnicyn Computing Centre, FRC CSC RAS, Russia
Alexander Kazakov	Matrosov Institute for System Dynamics and Control Theory, SB RAS, Russia
Oleg V. Khamisov L. A.	Melentiev Energy Systems Institute, Russia
Andrey Kibzun	Moscow Aviation Institute, Russia
Donghyun Kim	Kennesaw State University, USA
Roman Kolpakov	Moscow State University, Russia
Alexander Kononov	Sobolev Institute of Mathematics, Russia
Igor Konnov	Kazan Federal University, Russia

Vera Kovacevic-Vujcic	University of Belgrade, Serbia
Yury A. Kochetov	Sobolev Institute of Mathematics, Russia
Pavlo A. Krokhmal	University of Arizona, USA
Ilya Kurochkin	Institute for Information Transmission Problems, RAS, Russia
Dmitri E. Kvasov	University of Calabria, Italy
Alexander A. Lazarev	V.A. Trapeznikov Institute of Control Sciences, Russia
Vadim Levit	Ariel University, Israel
Bertrand M.T. Lin	National Chiao Tung University, Taiwan
Alexander V. Lotov	Dorodnicyn Computing Centre, FRC CSC RAS, Russia
Olga Masina	Yelets State University, Russia
Vladimir Mazalov	Institute of Applied Mathematical Research, Karelian Research Center, Russia
Nevena Mijajlović	University of Montenegro, Montenegro
Mikhail Myagkov	University of Oregon, USA
Angelia Nedich	University of Illinois at Urbana Champaign, USA
Yuri Nesterov	CORE, Université Catholique de Louvain, Belgium
Yuri Nikulin	University of Turku, Finland
Evgeni Nurminski	Far Eastern Federal University, Russia
Panos Pardalos	University of Florida, USA
Alexander V. Pesterev	V.A. Trapeznikov Institute of Control Sciences, Russia
Alexander Petunin	Ural Federal University, Russia
Stefan Pickl	Uni Bw Munich, Germany
Boris T. Polyak V.A.	Trapeznikov Institute of Control Sciences, Russia
Leonid Popov	IMM UB RAS, Russia
Mikhail A. Posypkin	Dorodnicyn Computing Centre, FRC CSC RAS, Russia
Alexander N. Prokopenya	Warsaw University of Life Sciences, Poland
Oleg Prokopyev	University of Pittsburgh, USA
Artem Pyatkin	Novosibirsk State University; Sobolev Institute of Mathematics, Russia
Radu Ioan Boţ	University of Vienna, Austria
Soumyendu Raha	Indian Institute of Science, India
Leonidas Sakalauskas	Institute of Mathematics and Informatics, Lithuania
Eugene Semenkin	Siberian State Aerospace University, Russia
Yaroslav D. Sergeyev	University of Calabria, Italy
Natalia Shakhlevich	University of Leeds, UK

Alexander A. Shananin	Moscow Institute of Physics and Technology, Russia
Angelo Sifaleras	University of Macedonia, Greece
Mathias Staudigl	Maastricht University, The Netherlands
Petro Stetsyuk	V.M. Glushkov Institute of Cybernetics, Ukraine
Fedor Stonyakin	V. I. Vernadsky Crimean Federal University, Russia
Alexander Strekalovskiy	Institute for System Dynamics and Control Theory, SB RAS, Russia
Vitaly Strusevich	University of Greenwich, UK
Michel Thera	University of Limoges, France
Tatiana Tchemisova	University of Aveiro, Portugal
Anna Tatarczak	Maria Curie-Skłodowska University, Poland
Alexey A. Tretyakov	Dorodnicyn Computing Centre, FRC CSC RAS, Russia
Stan Uryasev	University of Florida, USA
Frank Werner	Otto von Guericke University Magdeburg, Germany
Adrian Will	National Technological University, Argentina
Anatoly A. Zhigljavsky	Cardiff University, UK
Julius Žilinskas	Vilnius University, Lithuania
Yakov Zinder	University of Technology Sydney, Australia
Tatiana V. Zolotova	Financial University under the Government of the Russian Federation, Russia
Vladimir I. Zubov	Dorodnicyn Computing Centre, FRC CSC RAS, Russia
Anna V. Zykina	Omsk State Technical University, Russia

Organizing Committee Chairs

Milojica Jaćimović	Montenegrin Academy of Sciences and Arts, Montenegro
Yuri G. Evtushenko	Dorodnicyn Computing Centre, FRC CSC RAS, Russia
Nicholas N. Olenev	Dorodnicyn Computing Centre, FRC CSC RAS, Russia

Organizing Committee

Alla Albu	Dorodnicyn Computing Centre, FRC CSC RAS, Russia
Natalia Burova	Dorodnicyn Computing Centre, FRC CSC RAS, Russia

Alexander Gornov	Institute of System Dynamics and Control Theory, SB RAS, Russia
Vesna Dragović	Montenegrin Academy of Sciences and Arts, Montenegro
Vladimir Jaćimović	University of Montenegro, Montenegro
Michael Khachay	Krasovsky Institute of Mathematics and Mechanics, Russia
Yury Kochetov	Sobolev Institute of Mathematics, Russia
Vlasta Malkova	Dorodnicyn Computing Centre, FRC CSC RAS, Russia
Oleg Obradovic	University of Montenegro, Montenegro
Mikhail Posypkin	Dorodnicyn Computing Centre, FRC CSC RAS, Russia
Kirill Teymurazov	Dorodnicyn Computing Centre, FRC CSC RAS, Russia
Yulia Trusova	Dorodnicyn Computing Centre, FRC CSC RAS, Russia
Svetlana Vladimirova	Dorodnicyn Computing Centre, FRC CSC RAS, Russia
Ivetta Zonn	Dorodnicyn Computing Centre, FRC CSC RAS, Russia
Vladimir Zubov	Dorodnicyn Computing Centre, FRC CSC RAS, Russia

Contents

Mathematical Programming

A Derivative-Free Nonlinear Least Squares Solver

Igor Kaporin[✉][ORCID]

Federal Research Center "Computer Science and Control"
of the Russian Academy of Sciences, Vavilova 40, Moscow, Russia
igorkaporin@mail.ru

Abstract. An improved version of derivative-free nonlinear least squares iterative solver developed earlier by the author is described. First, we apply a regularization technique to stabilize the evaluation of search directions similar to the one used in the Levenberg-Marquardt methods. Second, we propose several modified designs for the rectangular preconditioning matrix, in particular a sparse adaptive techniques avoiding the use of pseudorandom sequences. The resulting algorithm is based on easily parallelizable computational kernels such as dense matrix factorizations and elementary vector operations thus having a potential for an efficient implementation on modern high-performance computers. Numerical results are presented for several standard test problems as well as for some special complex-valued cases to demonstrate the effectiveness of the proposed improvements to method.

Keywords: Nonlinear least squares · Derivative-free optimization · Pseudorandom preconditioning · Preconditioned subspace descent

1 Introduction

Application areas of nonlinear least squares are numerous and include, for instance, acceleration of neural network learning processes pattern recognition, signal processing etc. This explains the need in the further development of robust and efficient nonlinear least squares solvers.

The present paper continues the research started in [10] where a derivative-free version of the method [9] was presented. Similar to [3–5], we use the inexact Newton/Krylov subspace framework, however with search subspaces augmented by several previous directions, with different stepsize choice rule and with the use of rectangular preconditioner. The latter algorithimic feature (several variants of which are discussed below) is critically important for the nonlinear least square solver proposed in the present paper. Indeed, in the general case the residuals cannot be readily used to form search directions as it was done in [3–5] when the number of equations is equal to the number of unknowns.

© The Author(s), under exclusive license to Springer Nature Switzerland AG 2022
N. Olenev et al. (Eds.): OPTIMA 2022, CCIS 1739, pp. 3–17, 2022.
https://doi.org/10.1007/978-3-031-22990-9_1

2   General Description of Nonlinear LS Solver

A standard least squares problem is formulated as

$$x_* = \arg \min_{x \in R^n} \varphi(x), \tag{1}$$

where the function $\varphi : R^n \to R$ has the form

$$\varphi(x) = \frac{1}{2}\|f(x)\|^2 \equiv \frac{1}{2}f^\top(x)f(x), \tag{2}$$

and $f(x)$ is a nonlinear mapping $f : R^n \to R^m$, $m \geq n$. Assuming sufficient smoothness of f, an iterative procedure is constructed to find the minimizer x_* numerically. Note that x_* satisfies the equation

$$\operatorname{grad} \varphi(x_*) = 0, \tag{3}$$

where

$$\operatorname{grad} \varphi(x) = J^\top(x)f(x) \in R^n, \tag{4}$$

and

$$J(x) \equiv \frac{\partial f}{\partial x} \in R^{m \times n}, \tag{5}$$

is the Jacobian matrix of f at x. Despite of $m \geq n$, we still consider the zero residual case $\varphi(x_*) = 0$ which may have practical importance.

To make the exposition more self-contained, further we recall some results from [6–10].

2.1   Descent Along a Subnormalized Direction

Let $x_0, x_1, \ldots, x_t, \ldots$ be the sequence of approximations to the stationary point x_* constructed in the course of iterations, where t is the outer iteration index. Further on, we will use the notations

$$f_t = f(x_t), \quad J_t = J(x_t), \quad g_t = \operatorname{grad}(x_t) = J_t^\top f_t. \tag{6}$$

The next approximation x_{t+1} to x_* is constructed as

$$x_{t+1} = x_t + \alpha_t p_t, \tag{7}$$

where the stepsize parameter α_t satisfies

$$0 < \alpha_t < 2,$$

and p_t is a direction vector satisfying the subnormalization condition (as introduced in [10])

$$(J_t p_t)^\top (f_t + J_t p_t) \leq 0, \tag{8}$$

which can conveniently absorb the inexactness of the Jacobian by a vector products arising due to application of finite difference approximations. Note that the inequality (8) is a generalization of the normalization condition

$$(J_t p_t)^\top (f_t + J_t p_t) = 0$$

used earlier in [6–9], where the explicit availability of J_t as an $m \times n$ matrix was assumed. Next we consider sufficient conditions for the descent of $\varphi(x_t)$.

2.2   General Estimate for Residual Norm Reduction

Under rather mild conditions, see, e.g. [7–9] and appendix A below, there exists the limiting stepsize $\widehat{\alpha}_t \in (0, 2)$ such that for all $0 < \alpha \le \widehat{\alpha}_t$ the estimate

$$\frac{\varphi(x_t + \alpha p_t)}{\varphi(x_t)} \le 1 - \left(\left(\alpha - \frac{\alpha^2}{2} \right) \theta_t^2 \right)^2 \tag{9}$$

is valid, where φ is defined in (2), direction p_t is subnormalized by (8), and θ_t is determined as

$$\theta_t = \frac{\|J_t p_t\|}{\|f_t\|}. \tag{10}$$

Note that by the subnormalization condition (8) it holds

$$\vartheta_t = \vartheta(f_t, J_t p_t) \equiv \frac{-(J_t p_t)^\top f_t}{\|f_t\| \|J_t p_t\|} \ge \frac{\|J_t p_t\|}{\|f_t\|} = \theta_t, \tag{11}$$

so that the quantity (10) is a lower bound for the cosine ϑ_t of the Euclidean acute angle between m-vectors f_t and $(-J_t p_t)$. Clearly, estimate (9) shows the importance of finding subnormalized directions p_t with values of θ_t as large as possible. For the limiting stepsize condition sufficient for (9) to hold and the proof of the latter see Appendix A in the present paper.

Remark 1. It appears that $\widehat{\alpha}_t$ characterizes the nonlinearity of f in the neighborhood of x_t, while θ_t is related to the precision of approximate solution p_t of the "Newton equation" $f_t + J_t p_t = 0$. Note that the latter may not (and often cannot) be solved exactly in the context of our considerations.

2.3   Choosing the Value of the Stepsize

Based on estimate (9) one can develop the following Armijo type procedure [1] for evaluating appropriate stepsize α_t providing for a certain decrease of the residual norm. Let p_t be a direction vector satisfying the subnormalization condition (8). The value of stepsize is determined by checking the validity of estimate (23) (see Appendix below; we cannot directly use Jp as in earlier papers) for a decreasing sequence of trial values of $\alpha \in (0, 2)$; the standard choice is

$$\alpha^{(l)} = 2^{-l}, \qquad l = 0, 1, \ldots, l_{\max} - 1, \tag{12}$$

with $l_{\max} = 30$, which approximately corresponds to $\alpha^{(l)} > 10^{-9}$. As soon as (9) be satisfied (which is the case whenever $\widehat{\alpha}_t > 10^{-9}$), one sets $\alpha_t = \alpha^{(l)}$. Otherwise, the iterations are terminated with corresponding error message. However, in numerical testing, the backtracking criterion (9) was often satisfied at once for $l = 0$ with the stepsize $\alpha_t = 1$.

2.4   Approximating Product of Jacobian by a Vector

The derivative-free approximation for products like Jp is obtained using the second order central difference formula

$$J(x)p \approx \widetilde{J(x)p} = \frac{f(x + \zeta p) - f(x - \zeta p)}{2\zeta}, \tag{13}$$

where $\zeta = O(\tau^{1/3})$ and τ is the floating point tolerance. Assuming the sufficient smoothness of f, the precision of (13) is estimated as

$$\|J(x)p - \widetilde{J(x)p}\| = O(\tau^{2/3}),$$

which explains its further use. Indeed, in [10] it was noticed that under assumption (26) (see Appendix below), larger finite difference errors resulted from the use of the first order finite differences $J(x)p \approx (f(x + \zeta p) - f(x))/\zeta$ with $\zeta = O(\tau^{1/2})$ are incompatible with small values of θ defined by (10). (Relatively small θ_t can often be observed in computations; moreover, θ_t must decrease to zero as $t \to \infty$ for the convergence in nonzero residual problems.)

2.5   Choosing Subspace Basis and Descent Direction

Let $k \geq 1$ and $l \geq 0$ be fixed integers such that $k + l \leq n$. Recall that $p_{t-1} = (x_t - x_{t-1})/\alpha_{t-1}, \ldots$ are the previous search directions. Introduce the rectangular matrices $K_t \in R^{m \times n}$ serving as a kind of variable preconditioner, see Sect. 2.8 below. We further omit the iteration index t; for instance, we use notations

$$J = J(x_t), \qquad p_{k-i} = p_{t+k-i}, \qquad \widetilde{Ju_i} = \big(f(x_t + \zeta u_i) - f(x_t - \zeta u_i)\big)/(2\zeta).$$

According with numerical evidence presented in [10], the errors introduced by the use of finite difference approximations for the products of J by a vector are compensated by the application of Arnoldi-type orthogonalization procedure for constructing the bases of subspaces containing search directions:

$$v_1 \chi_{1,0} = f, \qquad v_{i+1} \chi_{i+1,i} = \widetilde{Ju_i} - \sum_{j=1}^{i} v_j \chi_{j,i}, \quad i = 1, \ldots, k + l,$$

where

$$u_i = K^\top v_i, \quad i = 1, \ldots, k,$$

$$u_i = p_{k-i}, \quad i = k + 1, \ldots, k + l,$$

and the coefficients $\chi_{1,0} = \|f\|$,

$$\chi_{j,i} = v_j^\top \widetilde{Ju_i}, \quad j = 1, \ldots, i, \qquad \chi_{i+1,i} = \Big\|\widetilde{Ju_i} - \sum_{j=1}^{i} v_j \chi_{j,i}\Big\|$$

are determined in a standard manner to satisfy the orthonormality condition $v_i^\top v_j = \delta_{i-j}$. The above procedure was proposed in [10] as a generalization of the ones presented in [7–9]. An equivalent form of the above recurrences is the matrix factorization

$$JU = VH + Z,$$ (14)

where

$$U = [u_1 \mid \ldots \mid u_{k+l}], \qquad V = [v_1 \mid \ldots \mid v_{k+l+1}], \qquad V^{\top}V = I_{k+l+1},$$

$$U \in R^{n \times (k+l)}, \quad V \in R^{m \times (k+l+1)}, \quad H \in R^{(k+l+1) \times (k+l)}, \quad Z \in R^{m \times (k+l)}.$$

Here, matrix Z accounts for the errors arising from the approximation of the Jacobian by a vector multiplications so that

$$z_i = Ju_i - \widetilde{Ju_i},$$

where the latter notation corresponds to (13). Therefore, the direction is determined as

$$p = Us, \quad s \in R^{k+l},$$ (15)

where s will be specified below in the next Section.

2.6 Characterizing Inexactness and Choosing Search Directions

Assume for a moment that $Z = 0$ in (14), which corresponds to exact computations with the Jacobian. In this case, the minimum norm solution of the form (15) for the overdetermined linear equation $f + Jp = 0$ is given by

$$s = -(H^{\top}H)^{-1}H^{\top}e_1\|f\|,$$

where $e_1 = [1 \ 0 \ \ldots \ 0]^{\top} \in R^{k+l+1}$ is the first unit vector. Indeed, according to the initialization of recurrences for v_i, one has

$$f = Ve_1\|f\|,$$

and, using (15) and (14) with $Z = 0$, it follows (recall that $V^{\top}V = I_{k+l+1}$)

$$\|f + Jp\| = \|Ve_1\|f\| + JUs\| = \|V(e_1\|f\| + Hs)\| = \|e_1\|f\| + Hs\|,$$ (16)

and the formula for the least squares solution readily follows.

As was announced in [10], a sufficiently reliable formula which takes into account both the presence of a nonzero Z and potential ill-conditioning of H is based on simple scaling and approximate pseudo-inversion

$$s = -\frac{1}{2}\left(H^{\top}H + \delta I\right)^{-1}H^{\top}e_1\|f\|, \qquad 0 < \delta \ll \|H\|_F^2,$$ (17)

where $\delta = \eta\|H\|_F^2$ and η is a small positive parameter. For the latter, a fixed value is used in actual implementation.

In order to estimate the effect of the derivative-free inexactness (which corresponds to $Z \neq 0$), the following condition seems to be the most convenient:

$$Z^{\top}Z \leq \xi^2(H^{\top}H + \delta I),$$ (18)

where ξ is a (typically small) positive parameter and δ is the same as in (17). Next we find some upper bounds for ξ under which search directions (17) possess the required subnormality (8) and angle (11) conditions.

2.7 Subnormality of Search Directions and the Lower Bound for θ

Introducing the notation

$$M = H(H^\top H + \delta I)^{-1} H^\top \in R^{(k+l+1)\times(k+l+1)}, \tag{19}$$

one finds

$$Hs = -\frac{1}{2} M e_1 \|f\|. \tag{20}$$

In the general case, similar to (16) one has

$$Jp = VHs + Zs, \qquad f + Jp = V(e_1\|f\| + Hs) + Zs,$$

and therefore

$$(Jp)^\top (f + Jp) = (s^\top H^\top V^\top + s^\top Z^\top)(V(e_1\|f\| + Hs) + Zs)$$

$$= s^\top H^\top (e_1\|f\| + Hs) + 2s^\top Z^\top V \left(\frac{1}{2} e_1\|f\| + Hs\right) + s^\top Z^\top Zs$$

$$= -\frac{1}{4}\|f\|^2 + \left\|\frac{1}{2} e_1\|f\| + Hs\right\|^2 + 2s^\top Z^\top V \left(\frac{1}{2} e_1\|f\| + Hs\right) + s^\top Z^\top Zs$$

$$\leq -\frac{1}{4}\|f\|^2 + \left\|\frac{1}{2} e_1\|f\| + Hs\right\|^2 + 2\|Zs\|\left\|\frac{1}{2} e_1\|f\| + Hs\right\| + \|Zs\|^2$$

$$= -\frac{1}{4}\|f\|^2 + \left(\left\|\frac{1}{2} e_1\|f\| + Hs\right\| + \|Zs\|\right)^2$$

$$\leq \frac{1}{4}\|f\|^2 \left(-1 + \left(\|(I - M)e_1\| + \xi\sqrt{e_1^\top M e_1}\right)^2\right)$$

$$\leq \frac{1}{4}\|f\|^2 \left(-1 + \left(\sqrt{1 - e_1^\top M e_1} + \xi\sqrt{e_1^\top M e_1}\right)^2\right)$$

where the last-but-one inequality holds by (17), (19), and (18) since

$$\|Zs\|^2 = s^\top Z^\top Zs \leq \xi^2 s^\top (H^\top H + \delta I)s = \frac{\xi^2}{4} e_1^\top M e_1 \|f\|^2,$$

while the last one follows from $M^2 \leq M$, so that

$$\|(I - M)e_1\| = \sqrt{1 - 2e_1^\top M e_1 + e_1^\top M^2 e_1} \leq \sqrt{1 - e_1^\top M e_1}.$$

Thus we obtain the estimate

$$(Jp)^\top (f + Jp) \leq \frac{1}{4}\|f\|^2 \left(-1 + \left(\sqrt{1 - e_1^\top M e_1} + \xi\sqrt{e_1^\top M e_1}\right)^2\right).$$

The required results can now readily be obtained.

First, the right hand side of the latter inequality is nonpositive if

$$\sqrt{1 - e_1^\top M e_1} + \xi \sqrt{e_1^\top M e_1} \le 1,$$

which gives us a simple sufficient condition

$$\xi \le \frac{\sqrt{e_1^\top M e_1}}{1 + \sqrt{1 - e_1^\top M e_1}},$$

for the subnormality (as defined by (8)) of direction (17).

Second, if one requires a more restrictive condition

$$\sqrt{1 - e_1^\top M e_1} + \xi \sqrt{e_1^\top M e_1} \le \rho \qquad (21)$$

for some $\rho \in (0,1)$ to be specified later, then two-side bounds for $\theta = \|Jp\|/\|f\|$ readily follow from the resulting inequality

$$(Jp)^\top (f + Jp) \le -\frac{1 - \rho^2}{4} \|f\|^2.$$

Indeed, from the latter inequality it follows

$$\|Jp\|^2 + \frac{1 - \rho^2}{4} \|f\|^2 \le -(Jp)^\top f \le \|Jp\|\|f\|$$

which is equivalent to the following quadratic inequality in θ,

$$\theta^2 - \theta + \frac{1 - \rho^2}{4} \le 0,$$

the solution of which is

$$\frac{1 - \rho}{2} \le \theta \le \frac{1 + \rho}{2}.$$

Setting now

$$\rho = 1 - \frac{1}{2}\sqrt{e_1^\top M e_1}$$

gives (quite similar to the one used in [10]) the estimate

$$\theta = \frac{\|Jp\|}{\|f\|} \ge \frac{1}{4}\sqrt{e_1^\top M e_1},$$

which is valid if

$$\xi \le -\frac{1}{2} + \frac{1 - \sqrt{1 - e_1^\top M e_1}}{\sqrt{e_1^\top M e_1}} \qquad (22)$$

according with (21). Note that the right hand side of (22) is positive whenever $e_1^\top M e_1 > 16/25$.

Remark 2. It only remains to notice that the quantity $e_1^\top M e_1 \in (0,1)$ monotonically increases in the progress of the Arnoldi iterations. Thus, the above presented condition (22) on ξ seems not to be too restrictive.

Therefore, the stepsize can be safely determined from appropriately modified estimate (9):

$$\frac{\|f(x + \alpha p)\|^2}{\|f\|^2} \leq 1 - \left(\left(\alpha - \frac{\alpha^2}{2}\right) \frac{e_1^\top M e_1}{16}\right)^2, \tag{23}$$

as soon as conditions (18) and (22) hold.

2.8 Using Quasirandom and Adaptive Rectangular Preconditioners

As the preconditioner we consider a full column rank matrix $K_t \in R^{m \times n}$ satisfying

$$K_t^\top K_t \approx I_n. \tag{24}$$

Clearly, the forming of K_t and multiplying it by a vector $q = K_t v$ must be as cheap as possible. Here we will consider preconditionings having a potential for a quite efficient implementation on modern high-performance computers. In the proposed designs, the so called logistic sequence (see, e.g., [19] and references cited therein) is used, defined as

$$\xi_0 \in (0, 0.5) \cup (0.5, 1), \qquad \xi_k = 1 - 2\xi_{k-1}^2, \qquad k = 1, 2, \ldots. \tag{25}$$

The idea of randomized preconditionings was already studied in [15] (see also the references cited therein), though in rather different context.

Quasirandom Hankel Matrix. As in [10], we consider K_t taken as $m \times n$ Hankel matrix $(K_t)_{i,j} = \xi_{i+j-1}$ with quasirandom entries generated by the logistic sequence.

Quasirandom Design with Hadamard Matrix. In this case, we consider K_t as the leading $m \times n$ submatrix of $H D_t H$, where D_t is a diagonal matrix with quasirandom entries generated by the logistic sequence, and H is the Hadamard matrix of the order $2^l \geq \max(m, n)$, defined recursively as

$$H_2 = \begin{bmatrix} 1 & 1 \\ 1 & -1 \end{bmatrix}, \quad H_{2k} = \begin{bmatrix} H_k & H_k \\ H_k & -H_k \end{bmatrix}, \quad k = 2, 4, 8, \ldots, 2^{l-1}.$$

Quasirandom Sparse Design. Here we used $K_t = [e_{k(1)} | \ldots | e_{k(n)}]$, where $e_k \in R^m$ is the kth unit vector, and the integer sequence $1 \leq k(1) < \ldots < k(n) \leq m$ was generated using quasirandom data associated with the logistic sequence (25). The procedure for evaluating $w = K^\top v$ has the following form:

$$k = 0;$$

for $i = 1, \ldots, m$:
    if $(|\xi_i| > \cos \frac{n\pi}{2m})$ **then**
       $k := k + 1$
       $w_k := v_i$
    end if
end for

Consequently, the number of nonzeroes in K_t is only n, condition (24) is satisfied exactly, and the computational costs for such preconditioning are minimal.

Adaptive Sparse Design. Here the preconditioner has the same form $K_t = [e_{k(1)} | \ldots | e_{k(n)}]$ as above, where $e_k \in R^m$ is the kth unit vector; however, in this case the integer sequence $1 \le k(1) < \ldots < k(n) \le m$ is constructed using the deterministic condition

$$|f_{k(i)}| \ge \frac{1}{2m} \sum_{j=1}^{n} |f_j|,$$

where f_j is the jth component of the current residual vector $f(x)$. An important additional advantage of such design is that it does not use any quasirandom data.

2.9   Description of Computational Algorithm

The above described preconditioned subspace descent algorithm can be summarized as follows. Note that indicating $f(x)$ as an input means the availability of computational module for the evaluation of vector $f(x)$ for any given x.

Algorithm 1.
Key notations:

$$V_t = [v_1 | \ldots | v_{i_{\max}+1}] \in R^{m \times (i_{\max}+1)}, \qquad U_t = [u_1 | \ldots | u_{i_{\max}}] \in R^{n \times i_{\max}},$$

$$H_i = \begin{bmatrix} \chi_{1,1} & \chi_{1,2} & \chi_{1,3} & \cdots & \chi_{1,i} \\ \chi_{2,1} & \chi_{2,2} & \chi_{2,3} & \cdots & \chi_{2,i} \\ 0 & \chi_{3,2} & \chi_{3,3} & \cdots & \chi_{3,i} \\ \cdots & \cdots & \cdots & \cdots & \cdots \\ 0 & \cdots & 0 & \chi_{i,i-1} & \chi_{i,i} \\ 0 & \cdots & 0 & 0 & \chi_{i+1,i} \end{bmatrix} \in R^{(i+1) \times i}, \qquad h_i = \begin{bmatrix} \chi_{1,1} \\ \chi_{1,2} \\ \chi_{1,3} \\ \cdots \\ \chi_{1,i} \end{bmatrix} \in R^i;$$

Input: $f(x) \in R^m$, $x_0 \in R^n$;
Initialization:
$s = k + l \le n$, $\eta = 10^{-12}$, $\zeta = 5 \cdot 10^{-6}$,
$\varepsilon = 10^{-10}$, $\tau_{\min} = 10^{-8}$, $t_{\max} = 10000$,
$f_0 = f(x_0)$, $\rho_0 = f_0^\top f_0$;
Iterations:
for $t = 0, 1, \ldots, t_{\max} - 1$:
    generate quasirandom $K_t \in R^{m \times n}$

$$v_1 := f_t/\sqrt{\rho_t}$$
$$w := v_1$$
$$i_{\max} := k + \min(l, t)$$
for $i = 1, \ldots, i_{\max}$:
$\qquad$ **if** $(i \leq k)$ **then**
$\qquad\qquad u_i := K_t^\top w$
$\qquad$ **end if**
$\qquad w := (f(x_t + \zeta u_i) - f(x_t - \zeta u_i))/(2\zeta)$
$\qquad$ **for** $j = 1, \ldots, i$:
$\qquad\qquad \chi_{j,i} = v_j^\top w$
$\qquad\qquad w := w - v_j \chi_{j,i}$
$\qquad$ **end for**
$\qquad \chi_{i+1,i} = \sqrt{w^\top w}$
$\qquad w := w/\chi_{i+1,i}$
$\qquad v_{i+1} = w$
end for
$$L_t L_t^\top = H_{i_{\max}}^\top H_{i_{\max}} + \eta \operatorname{trace}(H_{i_{\max}}^\top H_{i_{\max}})I$$
$$z_t := (L_t)^{-1} h_{i_{\max}}$$
$$\vartheta_t := z_t^\top z_t$$
$$z_t := (L_t)^{-\top} z_t \rho_t$$
$$p_t = -U_t z_t$$
$$u_{k+1+(t \bmod l)} := p_t$$
$$\alpha^{(0)} = 1$$
for $l = 0, 1, \ldots, l_{\max} - 1$:
$\qquad x_t^{(l)} = x_t + \alpha^{(l)} p_t$
$\qquad f_t^{(l)} = f(x_t^{(l)})$
$\qquad \rho_t^{(l)} = (f_t^{(l)})^\top f_t^{(l)}$
$\qquad \tau = \alpha^{(l)}(2 - \alpha^{(l)})\vartheta_t/16$
$\qquad$ **if** $(\tau < \tau_{\min})$ **return** x_t
$\qquad$ **if** $(\rho_t^{(l)}/\rho_t > 1 - (\tau/2)^2)$ **then**
$\qquad\qquad \alpha^{(l+1)} = \alpha^{(l)}/2$
$\qquad\qquad x_t^{(l+1)} = x_t + \alpha^{(l+1)} p_t$
$\qquad$ **else**
$\qquad\qquad$ **go to** NEXT
$\qquad$ **end if**
end for
NEXT: $x_{t+1} = x_t^{(l)}, \quad f_{t+1} = f_t^{(l)}, \quad \rho_{t+1} = \rho_t^{(l)};$
$\quad$ **if** $(\rho_{t+1} < \varepsilon^2 \rho_0)$ **or** $(\rho_{t+1} \geq \rho_t)$ **return** x_{t+1}
end for

Remark 3. The use of quantity ϑ_t can be explained as follows. To simplify the notation, let us drop the indices t, $i_{\max}$, and (l). Then, by $H^\top H + \delta I = LL^\top$, $h = H^\top e_1$, and (19), it holds

$$\vartheta = z^\top z = h^\top L^{-\top} L^{-1} h = e_1^\top H (H^\top H + \delta I)^{-1} H^\top e_1 = e_1^\top M e_1,$$

$$\frac{\tau}{2} = \frac{1}{2}\alpha(2-\alpha)\vartheta/16 = \left(\alpha - \frac{\alpha^2}{2}\right)\frac{e_1^\top M e_1}{16}.$$

Comparing the latter equality with (23) gives exactly the backtracking condition $\rho_t^{(l)}/\rho_t > 1 - (\tau/2)^2$ used in Algorithm 1 for the refinement of stepsize α.

3   Test Problems and Numerical Results

Below some results of application of Algorithm 1 to several standard hard-to-solve nonlinear test problems are presented. For the test runs, one core of Pentium(R) Dual-Core CPU E6600 3.06 GHz, 3.25 Gbytes RAM desktop PC was used. We will consider sufficiently large subspace dimensions with $k \geq l$ and $k + l \leq n$. For the nonzero residual problems, the iterations typically terminate by the condition $\tau < \tau_{\min} = 10^{-10}$, see the corresponding line in Algorithm 1.

3.1   Broyden Tridiagonal Function

Following [12], for $m = 500$ and $n = m$ define $f(x)$ as

$$f_i = (3 - 2x_i)x_i - x_{i-1} - 2x_{i+1} + 1, \quad 1 \leq i \leq 500,$$

where $x_0 = x_{n+1} = 0$. The optimum value is $f^\top f = 0$ and the starting point is set as $\widetilde{x} = [-1\ldots-1]^\top$. The results are presented in Table 1. This test can be considered as relatively easy due to the actual closeness of the initial guess $\widetilde{x}$ to the solution x_*.

Table 1. Broyden tridiagonal test: comparing various preconditionings

Preconditioning	$k+l$	#iter	#fun.eval.	opt.value	$\|x\|_C$
Quasirandom Hankel	50+50	15	1726	6.7E-10	0.707
Quasirandom Hadamard	50+50	14	1597	4.8E-10	0.707
Quasirandom sparse	50+50	5	526	3.7E-12	0.707
Adaptive sparse	50+50	8	864	1.9E-10	0.707

3.2   Chained Rosenbrock Function

This test function was introduced in [17], and we will use its version with $m = 2n - 2$ and essentially variable coefficients:

$$f_{2i-1} = i(x_i - x_{i+1}^2), \qquad f_{2i} = 1 - x_{i+1}, \qquad i = 1, \ldots, n-1.$$

The optimum value is $f^\top f = 0$ at $x_* = [1\ldots1]^\top$ and the starting point is $\widetilde{x} = [-1\ldots-1]^\top$. The results are given in Table 2 for $m = 198$ and $n = 100$. For this test case, the convergence history demonstrated the behavior typical for linear conjugate gradients with fast residual norm decrease at initial steps followed by a near stagnation phase and fast superlinear decrease at the final stage.

Table 2. Chained rosenbrock test: comparing various preconditionings

Preconditioning	$k+l$	#iter	#fun.eval	Opt.value	$\|x\|_C$
Quasirandom Hankel	50+50	664	133538	9.9E-08	1.000
Quasirandom Hadamard	50+50	721	145336	2.5E-08	1.000
Quasirandom sparse	50+50	632	126995	7.3E-08	1.000
Adaptive sparse	50+50	645	129379	8.2E-08	1.000

3.3 Approximate Canonical Decomposition of Inverse 3D Distance Tensor

This rather hard-to-solve nonzero residual problem was considered, e.g., in [11, 14,16]. Since the 3D array under consideration

$$t_{i,j,k} = \left(i^2 + j^2 + k^2\right)^{-1/2}$$

is symmetric, the residual function can be taken as

$$f_{i+(j-1)q+(k-1)q^2} = -\left(i^2 + j^2 + k^2\right)^{-1/2} + \sum_{l=1}^{r} x_{(l-1)q+i}x_{(l-1)q+j}x_{(l-1)q+k},$$

where $1 \le i,j,k \le q$, so that $m = q^3$ and $n = qr$. For $\tilde{x} = [1/2 \ldots 1/2]^\top$, the results are given in Table 3 for $r = 5$, $q = 30$ and in Table 4 for $r = 5$, $q = 50$. The sizes of the problem are $m = 27000$, $n = 150$ and $m = 125000$, $n = 250$.

Table 3. Inverse 3D distance tensor small test: comparing various preconditionings

Preconditioning	$k+l$	#iter	#fun.eval	Opt.value	$\|x\|_C$
Quasirandom Hankel	75+75	101	25008	0.01113926	0.601
Quasirandom Hadamard	75+75	76	17178	0.01113926	0.601
Quasirandom sparse	75+75	224	62676	0.01113926	0.601
Adaptive sparse	75+75	135	35149	0.01113926	0.601

Table 4. Inverse 3D distance tensor large test: comparing various preconditionings

Preconditioning	$k+l$	#iter	#fun.eval	Opt.value	$\|x\|_C$
Quasirandom Hankel	125+125	126	47880	0.02763617	0.653
Quasirandom Hadamard	125+125	102	36360	0.02763617	0.653
Quasirandom sparse	125+125	100	35480	0.02763617	0.653
Adaptive sparse	125+125	141	55278	0.02763617	0.653

3.4   Lennard-Jones Potential Minimization

The problem of finding

$$x = [r_1^\top \ldots r_N^\top]^\top = \arg\min_x \sum_{1 \le i < j \le N} \left(\|r_i - r_j\|^{-12} - 2\|r_i - r_j\|^{-6} \right),$$

where $r_i \in R^d$ and $d = 2$ or $d = 3$, serves as a popular hard-to-solve benchmark system for optimization algorithms, see, e.g., [2,13,18]. Its reformulation as a nonzero residual nonlinear LS problem with $m = N(N-1)/2$ and $n = dN$ readily follows if one sets

$$f_{i,j} = \|r_i - r_j\|^{-6} - 1, \qquad 1 \le i < j \le N.$$

Clearly, the minimum of the Lennard-Jones potential is expressed as

$$\min \sum_{1 \le i < j \le N} \left(\|r_i - r_j\|^{-12} - 2\|r_i - r_j\|^{-6} \right) = \min_x \|f(x)\|^2 - \frac{N(N-1)}{2}.$$

The results for $d = 2$, $N = 100$ (so that $m = 4950$ and $n = 200$) are shown in Table 5. The obtained minima well agree with that published in the existing literature: for 2D problem $f(x) = 68.26037$ yields -290.521 compared to -293.697 in [2]. For this problem (with rank deficient Jacobian and non-unique solution), dense quasirandom preconditionings provide for a considerably better results.

Note that for such complicated problems with multiple minima, the choice of the initial guess is probably the most important tuning parameter. In our tests with Lennard-Jones equations, we used 100 quasirandom initial guesses generated on the base of the logistic sequence as follows:

$$\xi_0 = 0.2, \qquad \xi_k = 1 - 2\xi_{k-1}^2, \qquad k = 1, 2, \ldots;$$

$$x_0^{(s)}(j) = \xi_{sj}/8, \qquad 1 \le j \le n, \qquad s = 1, 2, \ldots, 100;$$

the best results are shown in Table 5.

Table 5. Lennard-Jones 2D test: comparing various preconditionings

Preconditioning	$k + l$	#iter	#fun.eval	Opt.value	$\|x\|_C$
Quasirandom Hankel	180+20	74	29258	68.260376	7.915
Quasirandom Hadamard	180+20	63	24846	68.269600	10.23
Quasirandom sparse	180+20	89	35284	68.683881	50889
Adaptive sparse	180+20	99	39288	68.415685	1723

4   Concluding Remarks

In the present paper, a nonlinear least squares solver is presented which is based on derivative-free computations and is formally applicable to all types of least squares problems with sufficiently smooth residual function. Key feature of the algorithm is the use of quasirandom rectangular preconditioners for the construction of approximate Krylov subspaces containing descent directions. In practice, the proposed quasirandom Hadamard matrix based preconditioning can be recommended. The proposed algorithmic implementation of the method is justified by theoretical results related to the residual norm reduction. The results of numerical testing on several hard-to-solve problems have confirmed the efficiency and robustness of the derivatibe-free Preconditioned Subspace Descent method.

Acknowledgement. The author thanks the anonymous referee for insightful comments and suggestions which allow to significantly improve the exposition of the paper.

A   Limiting Stepsize Along Subnormalized Direction

Similar to [6,9], the proof of (9) is based on the assumption that the limiting stepsize $\widehat{\alpha} = \widehat{\alpha}(f,p)$ along a subnormalized direction p exists such that the limiting stepsize condition

$$\|f(x + \alpha p) - f - \alpha Jp\| \le \left(\alpha - \frac{\alpha^2}{2}\right) \frac{\|Jp\|^2}{\|f\|} \tag{26}$$

is satisfied for all $0 < \alpha \le \widehat{\alpha}$. (To clarify the notations, further we will omit the iteration index t.) For instance, a sufficient condition for (26) to hold is that f satisfies the local Lipschitz condition at x and $J(x)$ has full column rank, see, e.g., [9]. Indeed, (9) can be obtained from (26) and (8) as follows:

$$\|f(x + \alpha p)\| \le \|f + \alpha Jp\| + \|f(x + \alpha p) - f - \alpha Jp\|$$

$$= \left(\|f\|^2 + 2\alpha f^\top Jp + \alpha^2 \|Jp\|^2\right)^{1/2} + \|f(x + \alpha p) - f - \alpha Jp\|$$

$$\le \left(\|f\|^2 - 2\alpha\|Jp\|^2 + \alpha^2\|Jp\|^2\right)^{1/2} + \left(\alpha - \frac{\alpha^2}{2}\right) \frac{\|Jp\|^2}{\|f\|}$$

$$= \|f\| \left(\left(1 - (2\alpha - \alpha^2)\frac{\|Jp\|^2}{\|f\|^2}\right)^{1/2} + \left(\alpha - \frac{\alpha^2}{2}\right) \frac{\|Jp\|^2}{\|f\|^2} \right)$$

$$\le \|f\| \left(1 - \left(\left(\alpha - \frac{\alpha^2}{2}\right) \frac{\|Jp\|^2}{\|f\|^2}\right)^2\right)^{1/2},$$

where the latter estimate follows from the inequality

$$\sqrt{1 - \eta} + \frac{\eta}{2} \le \sqrt{1 - \frac{\eta^2}{4}},$$

which holds for any $0 \le \eta \le 1$ and is used with $\eta = \alpha(2 - \alpha)\|Jp\|^2/\|f\|^2$, see also (11).

References

1. Armijo, L.: Minimization of functions having Lipschitz continuous first partial derivatives. Pac. J. Math. **16**(1), 1–3 (1966)
2. Averick, B.M., Carter, R.G., Xue, G.L., More, J.J.: The MINPACK-2 test problem collection (No. ANL/MCS-TM-150-Rev.) Argonne National Lab., IL (United States) (1992)
3. Brown, P.N.: A local convergence theory for combined inexact-Newton/finite-difference projection methods. SIAM J. Numer. Anal. **24**(2), 407–434 (1987)
4. Brown, P.N., Saad, Y.: Hybrid Krylov methods for nonlinear systems of equations. SIAM J. Sci. Stat. Comput. **11**(3), 450–481 (1990)
5. Brown, P.N., Saad, Y.: Convergence theory of nonlinear Newton-Krylov algorithms. SIAM J. Optim. **4**(2), 297–330 (1994)
6. Kaporin, I.E.: Esimating global convergence of inexact Newton methods via limiting stepsize along normalized direction, report 9329. Department of Mathematics, Catholic University of Nijmegen, Nijmegen, The Netherlands, p. 8 (1993)
7. Kaporin, I.E.: The use of preconditioned Krylov subspaces in conjugate gradient type methods for the solution of nonlinear least square problems (Russian). Vestnik Mosk. Univ. Ser. (Computational Math. Cybernetics) **15**(3), 26–31 (1995)
8. Kaporin, I.E., Axelsson, O.: On a class of nonlinear equation solvers based on the residual norm reduction over a sequence of affine subspaces. SIAM J. Sci. Comput. **16**(1), 228–249 (1994)
9. Kaporin, I.: Preconditioned subspace descent method for nonlinear systems of equations. Open Comput. Sci. **10**(1), 71–81 (2020)
10. Kaporin, I.: A derivative-free nonlinear least squares solver. In: Olenev, N.N., Evtushenko, Y.G., Jaćimović, M., Khachay, M., Malkova, V. (eds.) OPTIMA 2021. LNCS, vol. 13078, pp. 217–230. Springer, Cham (2021). https://doi.org/10.1007/978-3-030-91059-4_16
11. Kazeev, V.A., Tyrtyshnikov, E.E.: Structure of the Hessian matrix and an economical implementation of Newton's method in the problem of canonical approximation of tensors. Comput. Math. Math. Phys. **50**(6), 927–945 (2010). https://doi.org/10.1134/S0965542510060011
12. More, J.J., Garbow, B.S., Hillstrom, K.E.: Testing unconstrained optimization software. Argonne National Laboratory, Applied Mathematics Division, Technical Memorandum No. 324, p. 96 (1978)
13. Northby, J.A.: Structure and binding of Lennard-Jones clusters: $13 \leq N \leq 147$. J. Chem. Phys. **87**(10), 6166–6177 (1987)
14. Oseledets, I.V., Savostyanov, D.V.: Minimization methods for approximating tensors and their comparison. Comput. Math. Math. Phys. **46**(10), 1641–1650 (2006). https://doi.org/10.1134/S0965542506100022
15. Pan, V.Y., Qian, G.: Randomized preprocessing of homogeneous linear systems of equations. Linear Algebra Appl. **432**, 3272–3318 (2010)
16. Sterck, H.D., Miller, K.: An adaptive algebraic multigrid algorithm for low-rank canonical tensor decomposition. SIAM J. Sci. Comput. **35**(1), B1–B24 (2013)
17. Toint, P.L.: Some numerical results using a sparse matrix updating formula in unconstrained optimization. Math. Comput. **32**(143), 839–851 (1978)
18. Wales, D.J., Doye, J.P.K.: Global optimization by basin-hopping and the lowest energy structures of Lennard-Jones clusters containing up to 110 atoms. J. Phys. Chem. A **101**(28), 5111–5116 (1997)
19. Yu, L., Barbot, J.P., Zheng, G., Sun, H.: Compressive sensing with chaotic sequence. IEEE Sig. Process. Lett. **17**(8), 731–734 (2010)

Gradient-Type Methods for Optimization Problems with Polyak-Łojasiewicz Condition: Early Stopping and Adaptivity to Inexactness Parameter

Ilya A. Kuruzov[1]([⊠]) [iD], Fedor S. Stonyakin[1,2] [iD],
and Mohammad S. Alkousa[1,3] [iD]

[1] Moscow Institute of Physics and Technology, Moscow, Russia
{kuruzov.ia,mohammad.alkousa}@phystech.edu, fedyor@mail.ru
[2] V. I. Vernadsky Crimean Federal University, Simferopol, Russia
[3] HSE University, Moscow, Russia

Abstract. Due to its applications in many different places in machine learning and other connected engineering applications, the problem of minimization of a smooth function that satisfies the Polyak-Łojasiewicz condition receives much attention from researchers. Recently, for this problem, the authors of [14] proposed an adaptive gradient-type method using an inexact gradient. The adaptivity took place only with respect to the Lipschitz constant of the gradient. In this paper, for problems with the Polyak-Łojasiewicz condition, we propose a full adaptive algorithm, which means that the adaptivity takes place with respect to the Lipschitz constant of the gradient and the level of the noise in the gradient. We provide a detailed analysis of the convergence of the proposed algorithm and an estimation of the distance from the starting point to the output point of the algorithm. Numerical experiments and comparisons are presented to illustrate the advantages of the proposed algorithm in some examples.

Keywords: Adaptive method · Gradient method · Polyak-Łojasiewicz condition · Inexact gradient

1 Introduction

With the increase in the number of applications that can be modeled as large- or even huge-scale optimization problems (some of such applications arise in machine learning, deep learning, data science, control, signal processing, statistics, and so on), first-order methods, which require low iteration cost as well as low memory storage, have received much interest over the past few decades

---

The research was supported by Russian Science Foundation and Moscow (project No. 22-21-20065, https://rscf.ru/project/22-21-20065/).

N. Olenev et al. (Eds.): OPTIMA 2022, CCIS 1739, pp. 18–32, 2022.
https://doi.org/10.1007/978-3-031-22990-9_2

[1]. Gradient-type methods may be regarded as the cornerstone and core of the numerical methods for solving optimization problems.

For the problem of minimization of a smooth function f, it is well known that if f is strongly-convex, then the gradient descent method achieves a global linear convergence rate [13]. However, many of the fundamental models in machine learning such as least squares and logistic regression yield objective functions that are convex but not strongly convex. This matter led to the formation of motivation for seeking some alternatives to strong convexity, showing that it is possible to obtain linear convergence rates for problems such as least squares and logistic regression. One of these alternatives is the Polyak-Łojasiewicz inequality (or PL-condition). This inequality was originally introduced by B. Polyak [15], who proved that it is sufficient to show the global linear convergence rate for the gradient descent without assuming convexity. PL-condition is very well studied by many researchers in many different works for many different settings of optimization problems and has been theoretically verified for objective functions of optimization problems arising in many practical problems. For example, it has been proven to be true for objectives of over-parameterized deep networks [2], learning LQR models [6], and phase retrieval [19]. More discussions of PL-condition and many other simple problems can be found in [9]. Note that many other important classes of non-convex problems (Lipschitz problems, weakly convex problems, weakly α-quasiconvex functions) are investigated by different authors (see e.g. [16,18,19]).

In the first-order methods (thus the classical gradient descent method), the availability of an exact first-order oracle is assumed. That is, the oracle must provide at each given point the exact values of the function and its gradient. But unfortunately, in many applications, there is no access to this exact information (especially information about the gradient) at each iteration of the method. This led researchers to investigate the behavior of first-order methods which have the possibility to work with an inexact oracle. This problem attracted the attention of many researchers in mathematical optimization. In [3] (which can be considered as a fundamental work in this direction) the authors introduce the notion of an inexact first-order oracle, which naturally appears in the context of smoothing techniques, Augmented Lagrangians, and many other situations. See [3–5,10,17,21] and references therein, for more details.

It is known that the analysis of the convergence of the gradient descent method, implied the constant step-size which depends on the Lipschitz constant of the gradient of the objective function (constant of smoothness). But in many applied optimization problems, it is difficult to estimate this constant. For example, the well-known Rosenbrock function and its multidimensional generalizations (for example, the Nesterov-Skokov function [12]) have only a locally Lipschitz continuous gradient. Thus, it is impossible to estimate the Lipschitz constant of the gradient for these functions without additional restrictions on the operation region of the method. In order to overcome the difficulty to estimate the value of the Lipschitz constant of the gradient, there have been many attempts to construct a method with adaptivity on the step-size, see for example [7,16–18].

Recently, for the problem of minimizing a smooth function that satisfies the PL-condition (which is the problem under consideration in this paper), in [14] the authors proposed non-adaptive and adaptive gradient-type methods using the notion of the non-exact gradient. They analyzed the proposed algorithms and the influence of non-exactness on the rate of convergence. But the adaptivity takes place with respect to the Lipschitz constant of the gradient, where it is still necessary to calculate the level of the noise in the gradient exactly.

In this paper, we continue the research in order to construct an adaptive gradient-type method that was studied in [14] for the first time and propose an adaptive algorithm for problems with objective functions that satisfy PL condition with a detailed analysis of its convergence and an estimation of the distance from the starting point to the output point of the algorithm. The adaptivity in the proposed algorithm in this paper will be in both parameters: the Lipschitz constant of the gradient and the level of the noise in the gradient. Therefore, the proposed algorithm is fully adaptive.

This paper consists of an introduction and 4 main sections. In Sect. 2 we formulate basic concepts, definitions, and assumptions that are connected with the problem under consideration. In Sect. 3 we mention an adaptive algorithm that was proposed in [14], the adaptivity in this algorithm is for the Lipschitz constant of the gradient of the objective function. Section 4 is devoted to a fully adaptive algorithm (the adaptivity in the Lipschitz constant of the gradient and the level of the noise in the gradient) for problems with objective functions that satisfy PL-condition. The last Sect. 5 is devoted to the numerical experiments which demonstrate the effectiveness of the proposed algorithm. The conducted experiments were conducted for the minimization problem of the quadratic form, the logistic regression problem, and for one minimization problem connected with the system of nonlinear equations.

2   Problem Statement and Basic Definitions

In this section, we present the problem statement and some basic concepts and definitions.

Definition 1. *The differentiable function* $f : \mathbb{R}^n \longrightarrow \mathbb{R}$ *is an L-smooth (or* ∇f *is Lipschitz-continuous) w.r.t.* $\| \cdot \|$, *for some constant* $L > 0$, *if*

$$\|\nabla f(x) - \nabla f(y)\| \leqslant L \|x - y\| \quad \forall x, y \in \mathbb{R}^n. \tag{1}$$

Here and everywhere in the paper, the norm $\| \cdot \|$ *indicates the Euclidean norm.*

Definition 2. *Let* f *be an L-smooth function. The gradient* ∇f *satisfies the Polyak-Łojasiewicz condition (for brevity, we write the PL-condition) [15] if the following inequality holds*

$$f(x) - f^* \leqslant \frac{1}{2\mu} \|\nabla f(x)\|^2 \quad \forall x \in \mathbb{R}^n, \tag{2}$$

where $\mu > 0$, $f^* = f(x_*)$ and x_* is one of the exact solutions of the optimization problem under consideration.

In this paper, we will consider a classical optimization problem

$$\min_{x \in \mathbb{R}^n} f(x), \tag{3}$$

when the objective function f satisfies (1) and (2).

Let us denote by $\widetilde{\nabla} f(x)$ the approximate value of $\nabla f(x)$ at any requested point x (alternatively we also call $\widetilde{\nabla} f(x)$ an inexact gradient of f at x), this means

$$\nabla f(x) = \widetilde{\nabla} f(x) + v(x), \quad \text{and} \quad \|v(x)\| \leqslant \Delta, \tag{4}$$

for some fixed $\Delta > 0$. From this, (2) means

$$f(x) - f^* \leqslant \frac{1}{\mu}(\|\widetilde{\nabla} f(x)\|^2 + \Delta^2) \quad \forall x \in \mathbb{R}^n. \tag{5}$$

3   Gradient Descent with an Adaptive Step-Size Policy

Due to the difficulty of estimating the Lipschitz constant of the gradient of the objective function, the researchers are actively working in order to propose methods that overcome this difficulty in adaptive forms. In [14], the authors proposed an adaptive algorithm (listed as Algorithm 1, below), which is a generalization of the universal gradient method [11] for working with an inexact gradient of the functions satisfying the PL-condition. The importance of working with inexact gradients occurs in many optimization problems in the Hilbert space [20] and, in a particular case, inverse problems [8]. The adaptivity in Algorithm 1, is for the Lipschitz constant L.

For the inexact gradient (4) we can get an inequality similar to (1), as follows:

$$f(x) \leqslant f(y) + \langle \widetilde{\nabla} f(y), x - y \rangle + L\|x - y\|^2 + \frac{\Delta^2}{2L}, \quad \forall x, y \in \mathbb{R}^n.$$

This inequality contains an exact calculation of the value of the function f at an arbitrary point from the domain of definition.

Let us assume that we can calculate the inexact value $\widetilde{f}$ of the function f at any point x, so that

$$|f(x) - \widetilde{f}(x)| \leqslant \delta, \tag{6}$$

for some $\delta > 0$. Then the following inequality holds:

$$\widetilde{f}(x) \leqslant \widetilde{f}(y) + \langle \widetilde{\nabla} f(y), x - y \rangle + L\|x - y\|^2 + \frac{\Delta^2}{2L} + 2\delta, \quad \forall x, y \in \mathbb{R}^n. \tag{7}$$

Algorithm 1. Adaptive Gradient Descent with Inexact Gradient [14].

Require: x_0, $L_{\min} \geqslant 0, L_0 \geqslant L_{\min}, \delta \geqslant 0, \Delta \geqslant 0$.

1: Set $k := 0$
2: Calculate

$$x_{k+1} = x_k - \frac{1}{2L_k}\widetilde{\nabla}f(x_k). \tag{8}$$

3: If

$$\widetilde{f}(x_{k+1}) \leqslant \widetilde{f}(x_k) + \langle \widetilde{\nabla}f(x_k), x_{k+1} - x_k \rangle + L_k\|x_{k+1} - x_k\|^2 + \frac{\Delta^2}{2L_k} + 2\delta, \tag{9}$$

then $k := k+1$, $L_k := \max\left\{\frac{L_{k-1}}{2}, L_{\min}\right\}$ and go to Step 2. Otherwise, set $L_k := 2L_k$ and go to Step 3.
4: **return** x_k.

For Algorithm 1, with a sufficiently small inexact gradient, at each point in the sequence $\{x_k\}_{k\geqslant 0}$,

$$\|\widetilde{\nabla}f(x_k)\| \leqslant 2\Delta, \tag{10}$$

and according to (5), we can guarantee that $f(x_k) - f^* \leqslant \frac{5\Delta^2}{\mu}$.

For Algorithm 1, in [14], the following theorem has been proved.

Theorem 1. *Suppose that f satisfies PL-condition (2) and conditions (6), $\Delta^2 \geqslant 16L\delta$ hold. Let the parameter $L_{\min}$ in Algorithm 1 be such that $L_{\min} \geqslant \frac{\mu}{4}$ and one of the following holds:*

1. *Algorithm 1 works N_* steps where N_* is such that*

$$N_* = \left\lceil \frac{8L}{\mu} \log\left(\frac{\mu(f(x_0) - f^*)}{\Delta^2}\right)\right\rceil. \tag{11}$$

2. *For some $N \leqslant N_*$, at the N-th iteration of Algorithm 1, stopping criterion (10) is satisfied for the first time.*

Then for the output point $\widehat{x}$ ($\widehat{x} = x_N$ or $\widehat{x} = x_{N_*}$) of Algorithm 1 will satisfy the following inequalities

$$f(\widehat{x}) - f^* \leqslant \frac{5\Delta^2}{\mu},$$

$$\|\widehat{x} - x_0\| \leqslant \frac{8\Delta}{\mu}\sqrt{\frac{\gamma^2}{2} + \frac{4\gamma L}{\mu}}\log\left(\frac{\mu(f(x_0) - f^*)}{\Delta^2}\right) + \frac{16\sqrt{\gamma L(f(x_0) - f^*)}}{\mu}, \tag{12}$$

where $\gamma = \frac{L}{L_{\min}}$. Also, the total number of calls to the subroutine for calculating inexact values of the objective function and (8) is not more than $2N + \log\frac{2L}{L_0}$.

Note, that Algorithm 1 uses subroutines for finding the inexact value of the objective function more often than the gradient method with a constant step-size. But the number of calls to these subroutines in Algorithm 1 is not more than $2N + \log \frac{2L}{L_0}$. This means that the "cost" of an iteration of the adaptive Algorithm 1 is, on average, comparable to about two iterations of the non-adaptive method (i.e. with a constant step-size). At the same time, the accuracy achieved by Algorithm 1 and the non-adaptive one also equals approximately [14].

4 Gradient Descent with Adaptivity in the Step-Size and Inexactness of the Noise Level

As we saw in the previous section, the adaptivity in Algorithm 1 was for the parameter L only. In this section, we consider an Algorithm (listed as Algorithm 2, below), which is a generalization of Algorithm 1 for the case of an unknown noise level Δ_k. It means that the adaptivity in Algorithm 2, will be for two parameters L and Δ. In this algorithm, at each iteration, we select the constants Δ_k and L_k, at each iteration, such that they satisfy the inequality for smooth functions with an inexact gradient for points from neighboring iterations.

Algorithm 2. Adaptive Gradient Descent Method for unknown L and Δ.

Require: $x_0, L_{\min} \geq \frac{\mu}{4} > 0, L_0 \geq L_{\min}, \Delta_0 > 0, \Delta_{\min} > 0.$
1: Set $k := 0$.
2: Set $L_k := \max\left\{\frac{L_{k-1}}{2}, L_{\min}\right\}$. for $k \geq 1$
3: Calculate:

$$x_{k+1} = x_k - \frac{1}{2L_k}\tilde{\nabla}f(x_k).$$

4: If

$$f(x_{k+1}) \leqslant f(x_k) + \langle \tilde{\nabla}f(x_k), x_{k+1} - x_k \rangle + \Delta_k\|x_{k+1} - x_k\| + \frac{L_k}{2}\|x_{k+1} - x_k\|^2,$$
(13)

then go to Step 5. Otherwise, set $L_k := 2L_k$ and $\Delta_k = 2\Delta_k$ and go to step 3.
5: Find the minimal value of Δ_k, such that (13) holds, and $\Delta_k \geq \Delta_{\min}$, also $\Delta_k \geqslant \Delta_j$, for $j < k$.
6: If stop condition does not hold, set $k := k + 1$ and go to Step 2.
7: **return** x_k.

For the sequence of points generated by Algorithm 2, due to fulfillment of condition (13), the following inequality holds

$$f(x_{k+1}) - f(x_k) \leqslant \frac{\Delta_k^2}{2L_k} - \frac{1}{4L_k}\|\tilde{\nabla}f(x_k)\|^2.$$
(14)

In this case, using (14), we can get the following estimate

$$f(x_{k+1}) - f^* \leqslant \prod_{j=0}^{k} \left(1 - \frac{\mu}{4L_j}\right)(f(x_0) - f^*) + \frac{\max_{j \leqslant k} \Delta_j^2}{\mu}, \qquad (15)$$

or by $L_j \leqslant \max_{j \leqslant k} L_j$, we get

$$f(x_{k+1}) - f^* \leqslant \left(1 - \frac{\mu}{4 \max_{j \leqslant k} L_j}\right)^{k+1}(f(x_0) - f^*) + \frac{\max_{j \leqslant k} \Delta_j^2}{\mu}. \qquad (16)$$

Here and below, by $\max_{j \leqslant k} \Delta_j^2$ we mean the maximum of all estimates Δ_j up to the k-th iteration.

Let us estimate the value $\max_{j \leqslant k} L_j$. Note that inequality (13) holds for $L_k \geqslant L$ and for $\Delta_k \geqslant \Delta$ for all $k \geqslant 0$.

Let us consider an arbitrary K-th iteration. If $\frac{\Delta}{\Delta_K} \leqslant \frac{L}{L_K}$, then when L_K reaches L, Δ_K reaches Δ also, it will not be more iterations in the step 4. Hence, if $\frac{\Delta}{\Delta_K} \leqslant \frac{L}{L_K}$, then $L_K \leqslant 2L$. On the other hand, if $\frac{\Delta}{\Delta_K} > \frac{L}{L_K}$, then in the worst case $L_K = L$ at the beginning and after the completion of the process, we have $L_K \leqslant \frac{2\Delta}{\Delta_K} L \leqslant \frac{2\Delta}{\Delta_{\min}} L$. Thus, we obtain the estimate $L_{\max} \leqslant 2L \max\left\{\frac{\Delta}{\Delta_{\min}}, 1\right\}$. Denoting $L_{\max} = L \max\left\{\frac{\Delta}{\Delta_{\min}}, 1\right\}$, we obtain the following refinement of the bound (16):

$$f(x_{k+1}) - f^* \leqslant \left(1 - \frac{\mu}{8L_{\max}}\right)^{k+1}(f(x_0) - f^*) + \frac{\max_{j \leqslant k} \Delta_j^2}{\mu}. \qquad (17)$$

In a similar way, for Δ_k we obtain that $\Delta_k \leqslant \Delta_{\max} := 2\Delta \max\left\{\frac{L}{L_{\min}}, 1\right\}$. Thus, $\max_{j \leqslant k} \Delta_j \leqslant \Delta_{\max}$.

In this case, we can stop the method after reaching the accuracy by the gradient $\|\tilde{\nabla} f(x_k)\| \leqslant 2\max_{j \leqslant k} \Delta_j$, which guarantees an estimate for the accuracy by the function

$$f(x_k) - f^* \leqslant \frac{5\max_{j \leqslant k} \Delta_j^2}{\mu} \leqslant \frac{5\Delta_{\max}^2}{\mu}.$$

On the other hand, using the estimate (17) and the introduced notation, we can guarantee the following rate of convergence

$$f(x_{k+1}) - f^* \leqslant \left(1 - \frac{\mu}{8L_{\max}}\right)^{k+1}(f(x_0) - f^*) + \frac{\Delta_{\max}^2}{\mu}. \qquad (18)$$

Then we get the following expression for the number of iterations

$$N \leqslant \left\lceil \frac{8L_{\max}}{\mu} \log\left(\frac{\mu(f(x_0) - f^*)}{4\Delta_{\max}^2}\right)\right\rceil.$$

As before, we get an estimate for the distance from the starting point to the current one, we can estimate it as follows:

$$\|x_N - x_0\| \leqslant \frac{8\Delta_{\max}}{\mu}\sqrt{\frac{\gamma^2}{2} + \frac{2\gamma L_{\max}}{\mu}\log\left(\frac{\mu(f(x_0) - f^*)}{4\Delta_{\max}^2}\right)} + \frac{16\sqrt{\gamma L_{\max}(f(x_0) - f^*)}}{\mu},$$

where $\gamma = \frac{L}{L_{\min}}$.

We can also estimate the total number of evaluations of the function f at each iteration of Algorithm 2. As mentioned earlier, the condition (13) is satisfied if $\Delta_k \geqslant \Delta$ and $L_k \geqslant L$. Thus, step 4 will be repeated at one iteration no more than

$$1 + \log_2 \left(\max \left\{ \frac{L}{L_{\min}}, \frac{\Delta}{\Delta_{\min}} \right\} \right).$$

Step 5 will be repeated no more than $\log_2 \left(\frac{2L}{L_{\min}} \right)$ times. Thus, the total number of function evaluations does not exceed

$$N_* \log_2 \left(\frac{4L}{L_{\min}} \max \left\{ \frac{L}{L_{\min}}, \frac{\Delta}{\Delta_{\min}} \right\} \right).$$

In this case, we get the following result about the work of Algorithm 2.

Theorem 2. *Let Algorithm 2 works either*

$$N_* = \left\lceil \frac{8L_{\max}}{\mu} \log \left(\frac{\mu(f(x_0) - f^*)}{4\Delta_{\max}^2} \right) \right\rceil$$

steps, or for some $k_* \leqslant N_*$ on the k_*-th iteration of Algorithm 2, the stopping criterion $\|\tilde{\nabla} f(x_{k_*})\| \leqslant 2 \max_{j \leqslant k} \Delta_j$ be achieved. Then for the output point $\widehat{x}$ ($\widehat{x} = x_{k_*}$ or $\widehat{x} = x_{N_*}$) of the Algorithm 2, the following inequality is guaranteed to be true

$$f(\widehat{x}) - f^* \leqslant \frac{5 \max_{j \leqslant k} \Delta_j^2}{\mu} \leqslant \frac{5\Delta_{\max}^2}{\mu}.$$

Moreover,

$$\|\widehat{x} - x_0\| \leqslant \frac{8\Delta_{\max}}{\mu} \sqrt{\frac{\gamma^2}{2} + \frac{2\gamma L_{\max}}{\mu} \log \left(\frac{\mu(f(x_0) - f^*)}{4\Delta_{\max}^2} \right)} + \frac{16\sqrt{\gamma L_{\max}(f(x_0) - f^*)}}{\mu},$$

where $\gamma = \frac{4L_{\max}}{\mu}$, $\Delta_{\max} = 2\Delta \max \left\{ \frac{L}{L_{\min}}, 1 \right\}$, $L_{\max} = L \max \left\{ \frac{\Delta}{\Delta_{\min}}, 1 \right\}$. Also, the total number of calls to the calculation of the function f is not more than

$$N_* \log_2 \left(\frac{4L}{L_{\min}} \max \left\{ \frac{L}{L_{\min}}, \frac{\Delta}{\Delta_{\min}} \right\} \right).$$

Remark 1. The value $L_{\max}$ estimates the maximum value of the parameter L_k. The estimates above remain valid if $L_{\max}$ is replaced by $\max_{j \leqslant k} L_j$.

Remark 2. Let at any point x we have a model $(\tilde{f}, \tilde{\nabla} f)$ of the function f such that the following conditions are satisfied:

$$\tilde{f}(x) \leqslant \tilde{f}(y) + \langle \tilde{\nabla} f(y), x - y \rangle + \frac{L}{2} \|x - y\|^2 + \Delta \|x - y\| + \delta, \qquad (19)$$

and

$$\tilde{f}(x) - f^* \leqslant \frac{1}{\mu} \left(\|\tilde{\nabla} f(x)\| + \Delta^2 \right) + \delta. \qquad (20)$$

Note that if $\tilde{\nabla} f$ is a Δ-inexact gradient and $\tilde{f}$ is a function such that $|f(x) - \tilde{f}(x)| \leqslant \delta$ for each x, then the conditions above are satisfied. A natural modification of the Algorithm 1 is the selection of L such that (19) will be satisfied for it. In this case, the achieved accuracy will decrease from $\frac{5\Delta^2}{\mu}$ to $\frac{5\Delta^2}{\mu} + \delta$. We can act similarly in the case of a known constant δ in Algorithm 2. In this case, the additional factor δ will also appear exactly. However, if the given parameter δ is not known, then it can also be selected. However, this will lead to an additional complication of the algorithm, which we will not discuss here. Note that the condition (13) will be hold if $L_k \geqslant L$ and $\Delta_k \geqslant \Delta + \frac{2L_k}{\|\tilde{\nabla} f(x_k)\|} \delta$. In order for this condition to be achieved, it is enough to modify the update of Δ_k at the 4th step of Algorithm 2 so that $\Delta_k := 2^\tau \Delta_k$ for $\tau > 1$. In this case, the estimates for the number of iterations and the accuracy of the solution with respect to the function, and the distance from the starting point to the output point from Theorem 2 will remain true, provided that the parameters $\Delta_{\max}$ and $L_{\max}$ will change accordingly.[1]

Remark 3. Note, that the achieved Δ_{max} can be significantly more than Δ according to results of Theorem 2. Nevertheless, note that we do not research in this work the influence of step 5 of Algorithm 2. Moreover, according to our experiments, the method stops when $\Delta_{\max} \sim \Delta$.

As before, these estimates give theoretical guarantees for the convergence of our method. But we observe that methods work significantly better in practice than in theory. Particularly, we see in all our experiments that proposed adaptive method Algorithm 2 can approach quality $O\left(\Delta^2\right)$ in terms of gradient norm like gradient method with constant step (see [14]).

5   Numerical Experiments

In this section, in order to demonstrate the performance of Algorithms 1, 2 and the algorithm with the constant step-size (see (2.1), (2.9) and Theorems 2, 3 in [14]) and compare them, we consider some numerical experiments concerning the quadratic optimization problem, logistic regression minimization and the solution of the system of nonlinear equations.

In all experiments, we will use an exact gradient with random noise $v(x)$ in (4), that is randomly generated, on the n dimensional sphere with radius 1 at the center 0, i.e. $v(x) \sim \mathcal{U}\left(S_1^n(0)\right)$.

---

[1] In the worst case, L_k at the beginning of a new iteration is already equal to L. Then denote by I_1 the minimal solution in I of the inequality $2^{\tau I} \Delta_{\min} \geqslant \Delta + \frac{\delta\sqrt{2}L}{\Delta} 2^I$ provided that $I \geqslant 1$. Then $L_{\max} = 2^I L$. Similarly, we can get that

$$\Delta_{\max} = 2\left(\Delta + \frac{\delta\sqrt{2}L_{\max}}{\Delta}\right) \cdot \max\left\{\left(\frac{L}{L_{\min}}\right)^\tau, 1\right\}.$$

All experiments were implemented in Python 3.4, on a computer fitted with Intel(R) Core(TM) i5-8250U CPU @ 1.60GHz, 8000 Mhz, 4 Core(s), 8 Logical Processor(s). The RAM of the computer is 8 GB.

Note, in the work [14] authors propose the stopping rule $\|\tilde{\nabla}f(x_k)\| \le \sqrt{6}\Delta$ for known inexactness value Δ. In this work constant $\sqrt{6}$ was chosen for a single criterion of stopping methods and we will use it too.

5.1 The Minimization Problem of the Quadratic Form

In this subsection, we consider the minimization problem of the quadratic form

$$f(x) = \sum_{i=1}^{n} d_i x_i^2, \quad x = (x_1,\ldots,x_n) \in \mathbb{R}^n, \, d_i \in \mathbb{R}. \quad (21)$$

We run Algorithms 1, 2 and the variant with a constant step-size (we denote this variant "Alg. constant" in the listed tables below), for $n = 100$, $L = \max_{1 \le i \le n} d_i = 1$ and different values of the parameter μ.

We take $x_0 = (100,\ldots,100)^\top$ as the initial point of all the compared algorithms.

The results are presented in Tables 1 and 2. The results, in Table 1 demonstrate the running time (in milliseconds) of the algorithms and the required number of iterations to achieve the accuracy $\|\tilde{\nabla}f(x)\| \le \sqrt{6}\Delta$, for different values of Δ. Meanwhile, the results in Table 2 demonstrate the achieved accuracy with respect to $\|\nabla f(x_N)\|$, which is the norm of the gradient of the objective function f at the output point x_N of the algorithms after N iterations, and the distance between the initial point x_0 and the output point x_N. Note that the distance between x_0 and the nearest optimal is equal to 948.7.

Table 1. The results of the algorithms for the quadratic form (21), with different values of μ and Δ, to achieve the accuracy $\|\tilde{\nabla}f(x)\| \le \sqrt{6}\Delta$.

μ	Δ	Alg. constant [14]		Algorithm 1		Algorithm 2	
		Iters	Time, ms	Iters	Time, ms	Iters	Time, ms
0.01	10^{-7}	1525	142.31	511	226.03	515	412.39
	10^{-4}	840	75.99	301	158.16	314	247.06
	10^{-1}	159	14.81	85	33.94	170	103.11
0.1	10^{-7}	169	15.96	76	29.58	102	60.81
	10^{-4}	104	10.15	49	18.11	94	44.58
	10^{-1}	41	4.46	24	8.21	54	32.44
0.9	10^{-7}	11	1.57	37	15.59	72	36.20
	10^{-4}	8	1.28	26	19.90	48	51.72
	10^{-1}	5	0.90	15	7.92	39	47.80
0.99	10^{-7}	6	1.02	34	13.38	58	34.01
	10^{-4}	5	0.89	24	9.21	46	27.20
	10^{-1}	3	0.52	14	5.06	48	24.74

Table 2. The results of the algorithms for the quadratic form (21), with different values of μ and Δ.

μ	Δ	Alg. constant [14]		Algorithm 1		Algorithm 2	
		$\|x_N - x_0\|$	$\frac{\|\nabla f(x_N)\|}{\Delta}$	$\|x_N - x_0\|$	$\frac{\|\nabla f(x_N)\|}{\Delta}$	$\|x_N - x_0\|$	$\frac{\|\nabla f(x_N)\|}{\Delta}$
0.01	10^{-7}	948.7	2.29	948.7	2.03	948.7	1.84
	10^{-4}	948.7	2.26	948.7	2.31	948.7	2.43
	10^{-1}	946.3	2.27	946.7	1.95	947.7	1.07
0.1	10^{-7}	948.7	2.17	948.7	1.97	948.7	0.87
	10^{-4}	948.7	2.18	948.7	1.68	948.7	0.80
	10^{-1}	948.3	2.14	948.3	2.26	948.5	0.83
0.9	10^{-7}	948.7	0.92	948.7	1.58	948.7	0.69
	10^{-4}	948.7	0.91	948.7	0.96	948.7	0.79
	10^{-1}	948.7	0.96	948.6	0.95	948.6	0.72
0.99	10^{-7}	948.7	0.96	948.7	0.95	948.7	0.71
	10^{-4}	948.7	0.95	948.7	0.92	948.7	0.68
	10^{-1}	948.7	1.05	948.7	0.93	948.6	0.69

We can see that the adaptive Algorithms 1 and 2 are slower than the non-adaptive one (algorithm with a constant step-size) for all parameters μ and Δ. However, they give a gain in the number of iterations for small μ. At the same time, we can notice that the running time for Algorithm 2 is longer than for Algorithm 1. The greatest difference is observed for large values of μ. At the same time, when μ decreases, they begin to show similar results. Also, we note from Table 2, that all three algorithms achieve approximately the same quality, while not going far from the starting point.

5.2   Logistic Regression

Now let us examine the work of the compared algorithms in the problem of logistic regression minimization, which has the following form

$$\min_{x \in \mathbb{R}^n} f(x) = \frac{1}{m} \sum_{i=1}^{m} \log\left(1 + \exp\left(-y_i \langle w_i, x \rangle\right)\right), \tag{22}$$

where $y = (y_1, \ldots, y_m)^\top \in [-1, 1]^m$ is the feasible variable vector, and $W = [w_1 \ldots w_m] \in \mathbb{R}^{n \times m}$ is the feature matrix, where the vector $w_i \in \mathbb{R}^n$ is from the same space as the optimized weight vector w.

Note that this problem may not have a finite solution in the general case. So we will create an artificial data set such that there is a finite vector x_* minimizing the given function in the way described in [14].

We chose $n = 200, m = 700$, and consider the case of constant inexactness. The results are presented in Table 3. From these results, we can see that the proposed Algorithm 2 stopped faster than gradient descent with constant and adaptive step-size for known Δ. But the obtained quality is a little worse. Nevertheless, we can see the main advantage of the proposed method, that it finds inexactness value without additional information.

Table 3. The results of Algorithms for the problem (22) with different values of Δ, to achieve the accuracy $\|\tilde{\nabla} f(x)\| \leqslant \sqrt{6}\Delta$.

Δ	Alg. constant [14]			Algorithm 1			Algorithm 2		
	Iters	Time, ms	$\frac{\|\nabla f(\tilde{x})\|}{\Delta}$	Iters	Time, ms	$\frac{\|\nabla f(\tilde{x})\|}{\Delta}$	Iters	Time, ms	$\frac{\|\nabla f(\tilde{x})\|}{\Delta}$
10^{-5}	20002	5604.34	3.56	902	2856.98	2.22	23	449.68	3.37
10^{-4}	9700	2605.02	2.42	472	1678.02	2.16	25	370.09	3.62
10^{-2}	83	49.86	2.29	17	68.36	2.10	17	161.27	0.84

5.3 Solving a System of Nonlinear Equations

In this subsection, we consider the problem of solving a system of m nonlinear equations

$$g_i(x) = \sum_{j=1}^{n} A_{ij} \sin(x_j) + B_{ij} \cos(x_j) := E_i, \quad i = 1, \ldots, m, \qquad (23)$$

where $x = (x_1, \ldots, x_n) \in \mathbb{R}^n$ and $A_{ij}, B_{ij} \in \mathbb{R}$, $\forall i = 1, \ldots, m; j = 1, \ldots, n$. This problem can be written as the following optimization problem

$$\min_{x \in \mathbb{R}^n} f(x) := \sum_{i=1}^{m} (g_i(x) - E_i)^2. \qquad (24)$$

In the conducted experiments, the parameters A_{ij}, B_{ij} were chosen such that $AB^\top = BA^\top = 0$, where A, B are the matrices with entries A_{ij}, B_{ij}, respectively. In this case, the jacobian of the system (23) will be $J = A \mathrm{diag}(\sin(x)) + B \mathrm{diag}(\cos(x))$. We can estimate the parameter μ so that

$$\mu = \lambda_{\min}(JJ^\top) \geqslant \min\left(\lambda_{\min}(AA^\top), \lambda_{\min}(BB^\top)\right),$$

where $\lambda_{\min}(X)$ is the smallest eigenvalue of the matrix X.

Also, the Lipschitz constant of the gradient of the objective function in (24), can be estimated so that $8\sqrt{2}\sigma_{\max}^2((A|B))$, where $\sigma_{\max}(X)$ is the largest singular value of the matrix X. Indeed, let $E \in \mathbb{R}^{m \times n}$ be a matrix formed by $E_i, i = 1, \ldots, m$, then the objective function can be represented as a composition $f(x) = h(\sin(x), \cos(x))$, where $h(x, y) = \|Ax + By - E\|^2$, with $L_h = 2\sigma_{\max}^2((A|B))$ as a Lipschitz constant of the gradient ∇h. Then for any $x, y \in \mathbb{R}^n$, we have

$$\|\nabla f(x) - \nabla f(y)\| \leqslant 4\|\nabla h(\sin(x), \cos(x)) - \nabla h(\sin(y), \cos(y))\|$$
$$\leqslant 4L_h \|(\sin(x) - \sin(y), \cos(x) - \cos(y))\|$$
$$\leqslant 4\sqrt{2}L_h \|x - y\|.$$

We run the compared algorithms for $n = 256$ and different numbers of equations $m \in \{8, 32, 128\}$. We take $x_0 = \mathbf{1}_n = (1, \ldots, 1)^\top$ as the initial point of all the compared algorithms. The results are presented in Tables 4 and 5. The results, in Table 4 demonstrate the running time (in milliseconds) of algorithms and the required number of iterations to achieve the accuracy $\|\tilde{\nabla} f(x)\| \leqslant \sqrt{6}\Delta$, for different values of Δ. Meanwhile, the results in Table 5, demonstrate the achieved accuracy with respect to the $\|\nabla f(x_N)\|$, which is the norm of the gradient of the objective function f at the output point x_N of the algorithms after N iterations, and the distance between the initial point x_0 and the output point x_N.

Table 4. The results of the algorithms for the problem (24), with different values of Δ, to achieve the accuracy $\|\tilde{\nabla} f(x)\| \leqslant \sqrt{6}\Delta$.

m	$\frac{L}{\mu}$	Δ	Alg. constant [14]		Algorithm 1		Algorithm 2	
			Iters	Time, ms	Iters	Time, ms	Iters	Time, ms
8	$2.1 \cdot 10^5$	10^{-4}	14434	2203.85	508	447.06	306	813.99
		10^{-1}	1479	220.82	56	45.54	69	178.63
32	$5.0 \cdot 10^6$	10^{-4}	59643	10314.68	1871	1808.70	510	1509.95
		10^{-1}	12536	2290.65	383	393.74	144	499.39
128	$7.7 \cdot 10^8$	10^{-4}	921805	177759.81	27405	23520.96	4688	11721.18
		10^{-1}	264361	47414.07	8800	7367.66	1916	4919.18

Table 5. The results of the algorithms for the problem (24), with different values of Δ.

m	$\frac{L}{\mu}$	Δ	Alg. constant [14]		Algorithm 1		Algorithm 2	
			$\|x_N - x_0\|$	$\frac{\|\nabla f(x_N)\|}{\Delta}$	$\|x_N - x_0\|$	$\frac{\|\nabla f(x_N)\|}{\Delta}$	$\|x_N - x_0\|$	$\frac{\|\nabla f(x_N)\|}{\Delta}$
8	$2.1 \cdot 10^5$	10^{-4}	5.1	2.41	5.1	2.25	5.1	3.07
		10^{-1}	4.8	2.34	4.8	2.09	4.9	2.28
32	$5.0 \cdot 10^6$	10^{-4}	7.4	2.41	7.5	2.27	7.5	3.31
		10^{-1}	7.4	2.37	7.4	2.24	7.4	2.72
128	$7.7 \cdot 10^8$	10^{-4}	14.4	2.43	14.4	2.33	14.4	15.67
		10^{-1}	14.3	2.43	14.3	2.32	14.4	1.84

Now, it can be seen from Table 4, that the adaptive Algorithms 1 and 2 converge much faster than the algorithm with a constant step-size. While the adaptive algorithms converge at approximately the same time, the algorithm with a constant step-size converges at least one and a half times slower. Moreover, for a large number $\frac{L}{\mu}$ it converges more than 10 times slower. A significant efficiency of the adaptive algorithms is observed for a large value of $\frac{L}{\mu}$ (see also the results in Table 5).

Note that, we stop the compared algorithms at accuracy $\|\tilde{\nabla} f(x)\| \leqslant \sqrt{6}\Delta$, although estimates for accuracy $\|\tilde{\nabla} f(x)\| \leqslant 2\Delta$ are proved for adaptive Algorithms 1 [14] and 2 (see Theorem 2). It was decided to choose a single stopping criterion that both adaptive algorithms can achieve the minimum accuracy from the available ones was chosen. If we consider the criterion $\|\tilde{\nabla} f(x_k)\| \leqslant \sqrt{6}\Delta$, then the results of Theorem 1 and 2 on the number of iterations and distances from x_0 to the output point $\hat{x} := x_N$ remain valid, but they can be refined by increasing the denominator in the logarithm.

6   Conclusion

In this paper, we have considered an adaptive gradient-type method for problems of minimizing a smooth function that satisfies the PL-condition. The adaptivity of the proposed algorithm is in the Lipschitz constant of the gradient and the level of the noise in the gradient. This gives the algorithm the attribute of being fully adaptive. A detailed analysis of its convergence, and an estimation of the distance from the starting point to the output point of the algorithm, were provided. Also, some numerical experiments were conducted for the problem of minimizing a quadratic form, logistic regression, and the problem of solving a system of nonlinear equations.

References

1. Beck, A.: First-Order Methods in Optimization. Society for Industrial and Applied Mathematics, Philadelphia (2017)
2. Du, S.S., Lee, J.D., Li, H., Wang, L., Zhai, X.: Gradient Descent Finds Global Minima of Deep Neural Networks (2018). https://arxiv.org/pdf/1811.03804.pdf
3. Devolder, O., Glineur, F., Nesterov, Yu.: First-order methods of smooth convex optimization with inexact oracle. Math. Program. **146**(1–2), 37–75 (2014)
4. Devolder, O.: Exactness, inexactness and stochasticity in first-order methods for largescale convex optimization: Ph.D. thesis (2013)
5. D'Aspremont, A.: Smooth optimization with approximate gradient. SIAM J. Opt. **19**(3), 1171–1183 (2008)
6. Fazel, M., Ge, R., Kakade, S., Mesbahi, M.: Global convergence of policy gradient methods for the linear quadratic regulator. In: Proceedings of the 35th International Conference on Machine Learning, PMLR 1980, Stockholm, Sweden, pp. 1466–1475 (2018)
7. Gasnikov, A.V.: Modern Numerical Optimization Methods. The Method of Universal Gradient Descent. A Textbook, 2nd edn. p. 272. MCCME (2021). (in Russian)
8. Kabanikhin, S.I.: Inverse and ill-posed problems: theory and applications. Walter de Gruyter, p. 475 (2011). https://doi.org/10.1515/9783110224016
9. Karimi, H., Nutini, J., Schmidt, M.: Linear convergence of gradient and proximal-gradient methods under the Polyak-Łojasiewicz condition. In: Frasconi, P., Landwehr, N., Manco, G., Vreeken, J. (eds.) ECML PKDD 2016. LNCS (LNAI), vol. 9851, pp. 795–811. Springer, Cham (2016). https://doi.org/10.1007/978-3-319-46128-1_50
10. Kuruzov, I.A., Stonyakin, F.S.: Sequential subspace optimization for quasar-convex optimization problems with inexact gradient. In: Olenev, N.N., Evtushenko, Y.G., Jaćimović, M., Khachay, M., Malkova, V. (eds.) OPTIMA 2021. CCIS, vol. 1514, pp. 19–33. Springer, Cham (2021). https://doi.org/10.1007/978-3-030-92711-0_2
11. Nesterov, Yu.: Universal gradient methods for convex optimization problems. Math. Program. **A**(152), 381–404 (2015)
12. Nesterov, Yu.E., Skokov, V.A.: Testing unconstrained minimization algorithms. In: The Book: "Computational Methods of Mathematical Programming". CEMI Ac. of Sc. M, pp. 77–91 (1980). (in Russian)
13. Nesterov, Yu.: Introductory Lectures on Convex Optimization. Springer Optimization and Its Applications, vol. 137. Springer, Heidelberg (2018)

14. Polyak, B.T., Kuruzov, I.A., Stonyakin, F.S.: Stopping Rules for Gradient Methods for Non-Convex Problems with Additive Noise in Gradient (2022). https://arxiv.org/pdf/2205.07544.pdf
15. Polyak, B.T.: Gradient methods for minimizing functionals. Comput. Math. Math. Phys. **3**(4), 864–878 (1963)
16. Sergeyev, Y.D., Candelieri, A., Kvasov, D.E., Perego, R.: Safe global optimization of expensive noisy black-box functions in the δ-Lipschitz framework. Soft. Comput. **24**(23), 17715–17735 (2020). https://doi.org/10.1007/s00500-020-05030-3
17. Stonyakin, F., et al.: Inexact relative smoothness and strong convexity for optimization and variational inequalities by inexact model. Optim. Methods Softw. **36**(6), 1155–1201 (2021)
18. Strongin, R.G., Sergeyev, Y.D.: Global Optimization with Non-convex Constraints: Sequential and Parallel Algorithms, vol. 45. Springer, Cham (2013)
19. Sun, J., Qu, Q., Wright, J.: A geometric analysis of phase retrieval. Found. Comput. Math. **18**(5), 1131–1198 (2018)
20. Vasilyev, F.: Optimization Methods. Fizmatlit, Moscow (2002). (in Russian)
21. Vasin, A., Gasnikov, A., Spokoiny, V.: Stopping rules for accelerated gradient methods with additive noise in gradient (2021). https://arxiv.org/pdf/2102.02921.pdf

Global Optimization

An Improved Genetic Algorithm for the Resource-Constrained Project Scheduling Problem

Evgenii N. Goncharov[1,2]([email icon]) [ORCID icon]

[1] Sobolev Institute of Mathematics, Prosp. Akad. Koptyuga, 4, Novosibirsk, Russia
[2] Novosibirsk State University, str. Pirogova, 1, Novosibirsk, Russia
gon@math.nsc.ru

Abstract. This paper presents an improved genetic algorithm for the Resource Constrained Project Scheduling Problem (RCPSP). The schedules are constructed using a heuristic that builds active schedules based on priorities that takes into account the degree of criticality for the resources. The degree of resource's criticality is derived from the solution of a relaxed problem with a constraint on accumulative resources. The computational results with instances from the PCPLIB library validate the effectiveness of the proposed algorithm. We have obtain some of the best average deviations of the solutions from the critical path value. The best known solutions have been improved for some instances from the PCPLIB.

Keywords: Project management · Resource-constrained project scheduling problem · Renewable resources · Genetic algorithms · PCPLIB

1 Introduction

The Resource-Constrained Project Scheduling Problem (RCPSP), denoted as $m, 1|cpm|C_{\max}$ [22], is one of the intractable optimization problems in operations research. RCPSP may be stated as follows: a project consists of a set activities V where each activity has to be processed without interruption. A partial order on the set of activities is defined by a directed acyclic graph. For every activity are assumed to be known duration, and the set and amounts of consumed resources. At every unit interval of the planning horizon $\hat{T}$ the same number of resources is allotted, and the resources are assumed to be unbounded outside the project horizon $\hat{T}$. All resources are renewable. The objective is to schedule the activities of a project to minimize the project makespan.

As a generalization of the job-shop scheduling problem the RCPSP is NP-hard in the strong sense [4]. So, it may be conceivable to use exact optimal

The study was carried out within the framework of the state contract of the Sobolev Institute of Mathematics (project FWNF-2022-0019).

© The Author(s), under exclusive license to Springer Nature Switzerland AG 2022
N. Olenev et al. (Eds.): OPTIMA 2022, CCIS 1739, pp. 35–47, 2022.
https://doi.org/10.1007/978-3-031-22990-9_3

methods only for projects of small size. For larger problems, one needs heuristics to get the best solution within a convenient response time, and heuristics remain the best way to solve these problems efficiently. Worth noting that introducing cumulative resources into the same problem makes the problem solvable with polynomial complexity [16].

The RCPSP is an important and challenging problem that has been widely studied over the past few decades. One of the most promising directions for developing heuristic methods is based on genetic algorithms (GA). Researchers have developed different schemes for representing solutions, genetic operations (crossover and mutation), and algorithms for solving RCPSP. We refer to the surveys provided by Pellerin at al. (2020) [36], Abdolshah (2014) [1], Vanhoucke (2012) [43], Hartmann and Briskorn (2010) [21], Kolisch and Hartmann (2006) [26].

In paper [18] author has proposed a genetic algorithm for the RCPSP problem. It has two crossovers, which creates an offspring according to the criterion of making the most use of available resources. Both crossovers use a heuristic rule to find promising segments (genes) of parent chromosomes for their further use in the offspring. This rule is based on the fact that first it turns out the "scarcity" of resources, which, in turn, we can get from solving a relaxed problem. To solve the relaxed problem, we use a known fast approximate algorithm [14]. In this paper, we propose a modified algorithm. It uses stochastics when choosing promising genes.

The quality of the proposed algorithm has been examined for instances of datasets j60, j90 and j120 from the electronic library PSPLIB [29]. We have found improved solutions for 5 instances from data set j60, and for 43 instances from data set j120. We have obtained some of the best average deviations from the critical path value. For some instances we have improved the best known literature solutions. We provide results of the numerical experiments.

2   Problem Setting

The RCPSP problem can be defined as follows. A project is taken as a directed acyclic graph $G = (N, A)$. Denote by $N = \{1, ..., n\} \cup \{0, n+1\}$ the set of activities in the project where activities 0 and $n + 1$ are dummy. The latter activities define the start and the completion of the project, respectively. The precedence relation on the set N is defined with a set of pairs $A = \{(i, j) \mid i \; precedes \; j\}$. If $(i, j) \in A$, then activity j cannot start before activity i has been completed. The set A contains all pairs $(0, j)$ and $(j, n + 1)$, $j = 1, ..., n$.

Denote by K the set of renewable resources. For each resource type $k \in K$, there is a constant availability $R_k \in Z^+$ throughout the project horizon $\hat{T}$. Activity j has deterministic duration $p_j \in Z^+$. The profile of resource consumption is assumed to be constant for every activity. So, activity j requires $r_{jk} \geq 0$ units of resource of type k, $k \in K$, at every time instant when it is processed. We assume that $r_{jk} \leq R_k$, $j \in N$, $k \in K$.

Let's introduce the problem variables. Denote by $s_j \geq 0$ the starting time of activity $j \in N$. Since activities are executed without preemptions, the completion time of activity j is equal to $c_j = s_j + p_j$. Define a schedule S as an $(n + 2)$-vector $(s_0, ..., s_{n+1})$. The completion time $T(S)$ of the project corresponds to the moment when the last activity $n + 1$ is completed, i.e., $T(S) = c_{n+1}$. Denote by $J(t) = \{j \in N \mid s_j < t \leq c_j\}$ the set of activities which are executed in the unit time interval $[t - 1, t)$ under schedule S. The problem is to find a feasible schedule $S = \{s_j\}$ respecting the resource and precedence constraints so that the completion time of the project is minimized. It can be formalized as follows: minimize the makespan of the project

$$T(S) = \max_{j \in N}(s_j + p_j) \tag{1}$$

under constraints

$$s_i + p_i \leq s_j, \quad \forall (i, j) \in A; \tag{2}$$

$$\sum_{j \in J(t)} r_{jk} \leq R_k, \ k \in K, \ t = 1, ..., \hat{T}; \tag{3}$$

$$s_j \in \mathbf{Z}^+, \quad j \in N. \tag{4}$$

Inequalities (2) define activities p recedence constraints. Relation (3) corresponds to the resource constraints and condition (4) define the variables in question.

3   Genetic Algorithm

Let's describe a modified GA algorithm (for more details of parent GA algorithm we refer to [18]). We use two crossover operators that are applied with equal probability during GA operation.

Solution Representation. We represent a feasible solution by the list of activities $L = (j_0, ..., j_{n+1})$ [27]. All lists considered are assumed to be compatible with the precedence relations. In other words, activity i is listed before activity j if $(i, j) \in A$. We use two algorithms for constructing schedules of activities with serial (S-SGS) and parallel (P-SGS) decoders. The serial decoding procedure [27] calculates an active schedule $S(L)$ for an arbitrary list L. A schedule is called active if none of the activities can be started earlier without delaying some other activity. It is known that there is an optimal schedule among active schedules. The parallel decoder (P-SGS) sequentially considers increasing schedule time, and determines all activities to be eligible which can be started up to the schedule time. We also apply a T-late decoder that uses the list L in reverse order and construct a so-called T-late schedule.

Initial Population. Each schedule for the initial population Γ is constructed as follows. We construct a random feasible list of activities L and then construct the schedule S with the parallel decoder. Further we apply a local improvement procedure FBI (the forward-backward improvement procedure) to the resulting schedule. The best schedule obtained in this way is added to the population. These actions are repeated until Γ contains the necessary number of schedules.

The Set of Parent Schedules. We look through the entire population Γ in nondecreasing order of schedule size, and add each schedule to the set of parents Γ' with a certain probability. If we have looked through the entire population and have not collected the necessary number of schedules $ParentsSize$ (an algorithm parameter), we choose the best among those which have not yet been added. A pair of parent schedules will be randomly selected for a crossover from the set Γ' of parent schedules. It has been noted in literature that this strategy for choosing a pair of parent schedules has proved to work well in problems of large dimension (see for example [37]).

We transmit to the offspring schedule the segment of the parent chromosome (gene) that most optimally uses available resources, i.e., yields the smallest surplus of resources. At each schedule time, we find the surplus of unused resources. If it is less than a predefined admissible residue, we define the set of activities fulfilled during this time interval. Let's call this set of activities a dense gene. In general case, we can operate with weighted residues of unused resources.

Heuristic Rule for Choosing Dense Genes. On the preliminary stage, before GA begin, we find the degree of scarcity for each resource and rank them, assigning them a weights. Let Θ be the set of dense genes and w_k be the weight of a resource of type k, $k = 1, ..., K$. We can compare the weights of genes, giving priority to those of them, which the higher priority (scarce) resources are used rationally, i.e., give less surplus of unused resources.

$DenseActivities(S, R) \rightarrow \Theta$.

1. Set $t := 0$, $\Theta = \emptyset$.
2. While $t < T(S)$:
   (a) find the weight v_t of the set $A(t)$: $v_t = \sum_{k \in K} \left(R_k - \sum_{j \in A(t)} r_{jk} \right) w_k / R_k$;
   (b) if $v_t < R$, then update the list of dense genes $\Theta = \Theta \cup A(t)$;
   (c) $t := t + 1$.

It may happen that some of dense genes intersect. In this case, we will leave the gene that yields smaller weighted surplus of unused resources v_t.

Resource Weights. We determine the degree of relative scarcity for the resources by solving a relaxed problem. For this purpose, we weaken the renewability condition for the resources and consider a problem with storable resources, that is, instead of constraints (3) we consider constraints

$$\sum_{t'=1}^{t} \sum_{j \in A(t')} r_{jk} \leq \sum_{t'=1}^{t} R_k, \quad k \in K, \ t = 1, ..., \hat{T}. \tag{5}$$

There is a known fast algorithm to solve problem (1), (2), (5), (4), whose computational complexity depends as a function of order n^2 on the number n of arcs in the reduction graph G. These algorithm is asymptotically exact with absolute error that tends to zero as the problem dimension grows in the case of real activity durations [14]. It is also known the exact algorithm where activity durations are integer [16]. Let us choose the first one. Applying this algorithm, we get, apart from a solution for the relaxed problem, the residues of storable resources. We define the degree of scarcity for the resources: a resource with a smaller balance of unused resource will be considered more scarce. We then apply the resulting resource ranking rule obtained in the relaxed problem to the original problem (1)–(4).

Unlike Algorithm [18], resource weights w_k are random variables. Let $i_1, \ldots i_k$ be a list of resources in ascending order of their ranks. Define $\gamma_{i_1} = 1 < \gamma_{i_2} < \ldots < \gamma_{i_k}$. Let's put now $w_{i_1} = 1$, w_{i_2} takes a random value from the interval $(\gamma_{i_1}, \gamma_{i_2})$, and so on, w_{i_k} is a random variable from the interval $(\gamma_{i_{k-1}}, \gamma_{i_k})$. We use uniform distribution to assign the random weights in the intervals. The idea of randomly assigning resource weights is that: we get more variety in choosing the best genes among dense ones. Besides, we can also vary the set of dense genes itself by weighting the genes. Numerical experiments confirmed our assumptions. Note, we can also vary parameters γ while the algorithm is running.

4   Crossovers and Algorithm Scheme

We use two crossovers presented in [18] without modification. These crossover procedures is based on finding a set of activities whose simultaneous fulfillment leads to small weights of the remaining unused resources. Let's describe them briefly.

Crossover Procedure A. Consider the first (in order) dense genes in each parent schedule and choose as the leading one gene that corresponds to the smaller weight. In the offspring chromosome, we place the activities from the beginning of the parent chromosome that contains the leading gene up to the last activity of the leading gene, inclusively. Next, consider the first dense genes in the schedules that do not contain activities already added to the daughter chromosome. Choose the best of the two new genes and add activities from the parent chromosome that contained it (except, obviously, the activities that have already been added to the daughter chromosome). Repeat these actions until we have considered all dense genes. In conclusion, if necessary, augment the daughter chromosome with remaining activities in the same order in which they are located in the schedule with minimal duration. We construct the daughter schedule by the daughter chromosome constructed in this way with the serial decoder.

Crossover algorithm A can be written as follows.
$CrossingA(S_1, S_2) \to S$.

1. Let daughter chromosome $L := \emptyset$.
2. Find dense genes for schedules $\Theta_1 := DenseActivities(S_1, R)$, $\Theta_2 := DenseActivities(S_2, R)$.
3. While $\Theta_1 \neq \emptyset$ & $\Theta_2 \neq \emptyset$, do:
   (a) if $v(D_{1,1}) \leq v(D_{2,1})$, then:
      i. find in chromosome L_1 position k of the last activity from gene $D_{1,1}$,

ii. add to L the activities from L_1, starting from j_1 and ending with j_k, except for the already added activities,

iii. remove from Θ_1 and Θ_2 genes that have been considered;

(b) otherwise apply step 3 to $D_{2,1}$.

4. If $\Theta_1 = \emptyset$ & $\Theta_2 \neq \emptyset$, then apply step 3 to $D_{2,1}$.
5. If $\Theta_1 \neq \emptyset$ & $\Theta_2 = \emptyset$, then apply step 3 to $D_{1,1}$.
6. If $\Theta_1 = \emptyset$ & $\Theta_2 = \emptyset$ & $|L| < |N|$, then add the missing activities in the order in which they come in the better schedule out of S_1 and S_2.
7. Construct schedule $S := SerialDecoder(L)$.

Crossover Procedure B. In this crossover, we construct a schedule from the neighborhoods $N_A(S)$ and $N_T(S)$ [18].

A block of activity j in an active schedule S is a set of activities that are overlapped to activity j in a given feasible schedule, started immediately after activity j, or finished immediately before it. Graph G_S is a digraph with set of vertices $V = N$ and set of arcs $E = \{(i,j) \mid c_i = s_j, (i,j) \in A\}$. The outgoing network of activity j for schedule S is the maximal (with respect to inclusion) connected subgraph of graph G_S where the only source is the vertex corresponding to activity j.

Choose the dense genes with the lowest weight criterion from the both parent schedules. Then, we reveal these activities in the second chromosome, we find the outgoing (or incoming) network for each of them and find the segment in the list of activities between the leftmost and rightmost activities in the block and outgoing (incoming) networks.

$CrossingB(S_1, S_2) \to S$.

1. Find dense genes for parent schedules
$\Theta_1 := DenseActivities(S_1, R)$, $\Theta_2 := DenseActivities(S_2, R)$.
2. Find gene $D \in \Theta_1 \cup \Theta_2$ of the minimal weight. Without loss of generality, we assume that $D \in L_1$.
3. Mark the segment $\alpha := D$ in L_2.
4. For each activity $j \in D$, add to α the activities from the network outgoing from j in L_2.
5. Find positions j_1 and j_2 in chromosome L_2 of the leftmost and rightmost activities from α.
6. Include into α all still unadded activities between j_1 and j_2 from chromosome L_2.
7. Construct the daughter schedule S:
(a) assign activities in positions up to and including the j_1th to the schedule as in S_2,
(b) assign activities between positions j_1 and j_2 to the schedule with the parallel decoder,
(c) assign activities after the j_2th to the schedule with the serial decoder.

The Mutation Operator. The mutation operator is to "mix" the activities in the list corresponding to the schedule and construct the schedule by the resulting

list of activities. Mutation is done in two stages. On the first stage, we choose
a pair of random activities in the chromosome and switch their places unless
it violates precedence conditions. On the second stage, we move one random
activity to a different place, while also not violating the precedence conditions.
This procedure is repeated a given number of times.

The Next Generation. We choose a given number (a parameter of the algorithm) of the best schedules from the set of offspring schedules Γ''. They are
added to the next generation Γ. The same number of the worst schedules are
removed from Γ.

The GA Algorithmn. Denote by λ the maximal number of generated schedules. The number of generated schedules will be increased for every computation
of the objective function, and exceeding parameter λ will be used as the stopping
criterion for the algorithm. The general scheme of the algorithm is as follows.

1. Create initial population $\Gamma := InitialPopulation(PopulationSize)$ and store
   the record S^*.
2. Initiate a set of offspring schedules $\Gamma'' := \emptyset$.
3. While the number of generated schedules does not exceed λ:
   (a) construct the set of parent schedules
       $\Gamma' = ParentsSelection(\Gamma, ParentsSize)$,
   (b) until the necessary number of offspring schedules has been generated, do:
       i. choose two parent schedules S_1 and $S_2 \in \Gamma'$ at random,
       ii. choose crossover $Crossing$ with equal probability from $CrossingA$
           and $CrossingB$,
       iii. cross S_1 and $S_2 : S' := Crossing(S_1, S_2)$,
       iv. apply to S' sequentially the operations of mutation and local improve-
           ment FBI: $S' := Mutation(S', MutationSize)$, $S' := FBI(S')$,
       v. if the mutation has made S' worse, it is canceled,
       vi. if $T(S') < T(S^*)$, then update the record $S^* := S'$,
       vii. update the set of offspring schedules $\Gamma'' = \Gamma'' \cup S'$,
   (c) create a new population for the next generation,
   (d) if S^* has not been updated for a given number of steps, replace a given
       number of the worst schedules in the population with new schedules.

5 Numerical Experiments

The GA algorithm was coded in C++ in the Visual Studio system and run on
a 3.4 GHz CPU and 16 Gb RAM computer under operating system Windows
7. In order to evaluate the performance of the proposed algorithm, we use the
standard set of instances presented in Kolisch and Sprecher [29]. These instances
are available in the project scheduling library PSPLIB along with their best-
known values. The instances are downloadable at http://www.om-db.wi.tum.
de/psplib/.

Optimal solutions for instances from the datasets j60, j90 and j120 are
unknown. The measure of the solution quality is the average percent deviation

Table 1. Average deviations from the critical path for dataset j60.

Algorithm	Reference	Year	APD, %	
			$\lambda = 50000$	$\lambda = 500000$
GA	this paper	2022	**10,50**	**10,40**
GA	Goncharov, Leonov [18]	2017	10,52	10,42
GANS	Proon, Jin [37]	2011	10,52	–
TS, VNS	Goncharov [20]	2022	10,55	10,44
Sequential(SS(FBI))	Berthaut et al. [3]	2018	10,58	10,45
GH + SS(LS)	Paraskevopoulos et al. [39]	2012	10,54	10,46
AI(FBI)	Mobini, at al. [34]	2011	10,55	–
TS + SS(FBI)	Mobini, at al. [33]	2009	10,57	–
GA(FBI)	Wang et al. [45]	2010	10,57	–
GA(FBI)	Goncalves [17]	2011	10,57	10,49
EA(GA(LS)+DEA(LS))	Elsayed et al. [11]	2017	10,58	–
PSO(LS)	Czogalla and Fink [8]	2009	10,62	–
GA	Lim et al. [31]	2013	10,63	10,51
Parallel(MA(LS))	Chen, at al. [7]	2014	10,63	–
PL(LS)	Zheng and Wang [49]	2015	10,64	–
GA(FBI)	Zamani [48]	2013	10,65	–
SFL(LS)	Fang and Wang [12]	2012	10,66	–
GA(FBI)	Ismail and Barghash [23]	2012	10,66	–
ACOSS	Wang Chen, at al. [44]	2010	10,67	–
GAPS	Mendes, at al. [32]	2009	10,67	10,67
GA	Debels, Vanhoucke [10]	2007	10,68	–
Specialist(PSO(LS))	Koulinas et al. [30]	2014	10,68	–
GA(LS)	Carlier et al. [6]	2009	10,70	–
Scatter search - FBI	Debels, et al. [9]	2006	10,71	10,53

Table 2. Average deviations from the critical path for dataset j90.

Algorithm	Reference	Year	APD, %	
			$\lambda = 50000$	$\lambda = 500000$
GA	this paper	2022	9,92	**9,61**
GA	Debels, Vanhoucke [10]	2017	**9,90**	–
Sequential(SS(FBI))	Berthaut et al. [3]	2018	9,96	9,74
TS, VNS	Goncharov [20]	2022	9,98	9,78
Sequential(SS)	Ranjbar and Kianfar [38]	2009	10,04	–
SS(EM + FBI)	Debels, et al. [9]	2006	10,09	9,80
PL(LS)	Jedrzejowicz, Ratajczak [24]	2006	11,60	–
TS	Ying et al. [47]	2009	12,15	–

Table 3. Average deviations from the critical path for dataset j120.

Algorithm	Reference	Year	APD, %	
			$\lambda = 50000$	$\lambda = 500000$
GA	this paper	2022	30,46	**29,63**
Specialist GA	Goncharov, Leonov [18]	2017	30,50	29,74
TS, VNS	Goncharov [19]	2019	30,56	29,88
GA	Lim et al. [31]	2013	30,66	29,91
biased random-key GA	Goncalves [17]	2011	32,76	30,08
GANS	Proon, Jin [37]	2011	**30,45**	30,78
ACOSS	Wang Chen, at al. [44]	2010	30,56	–
DBGA	Debels, Vanhoucke [10]	2007	30,69	–
GH + SS(LS)	Paraskevopoulos et al. [39]	2012	30,78	30,39
GA	Debels, Vanhoucke [10]	2007	30,82	–
PL(LS)	Zheng and Wang [49]	2015	31,02	–
SFL(LS)	Fang and Wang [12]	2012	31,11	–
Sequential(SS(FBI))	Berthaut et al. [3]	2018	31,16	30,39
EA(GA(LS)+DEA(LS))	Elsayed et al. [11]	2017	31,22	–
Specialist(PSO(LS))	Koulinas et al. [30]	2014	31,23	–
GA - Hybrid, FBI	Valls, at al. [41]	2008	31,24	30,95
GA(FBI)	Wang et al. [45]	2010	31,28	–
GA(FBI)	Zamani [48]	2013	31,30	–
Enhanced SS	Mobini, at al. [33]	2009	31,37	–
GA(LS)	Alcaraz and Maroto [2]	2006	31,38	–
GA(LS)	Carlier et al. [6]	2009	31,40	–
Scatter search - FBI	Debels, et al. [9]	2006	31,57	30,48
GAPS	Mendes, at al. [32]	2009	31,44	31,20
GA, FBI	Valls, et al. [40]	2005	31,58	–

(APD) of the received solutions from the lower bounds obtained by the critical path algorithm.

In Tables 1 – 3 we show comparison the GA algorithm performance with the previous results of experimental evaluation of competitive heuristics for the dataset j60, j90 and j120 respectively. The scrutiny of the presented results clearly shows the good performance of the proposed algorithm. We found improved solutions for 5 instances from the data set j60, and for 43 instances from the data set j120.

Average processing time is 16 s for $\lambda = 50000$ and 150 s for $\lambda = 500000$.

6   Conclusion

We have proposed a genetic algorithm for the resource-constrained project scheduling problem with respect to the makespan minimization criterion. We have developed two versions of the neighborhoods. The algorithm uses a heuristic that takes into account the degree of criticality (scarcity) of the resources, which is derived from the solution of the relaxed problem with constraints with cumulative resources. We have conducted numerical experiments on sets of instances from the PSPLIB electronic library. The results of the computational experiments suggest that the proposed algorithm is a very competitive heuristic and yields better results than several heuristics presented in the literature. The best known heuristic solutions have been improved for some instances from the dataset j120.

Further studies will be focused on constructing hybrid algorithms for the RCPSP problem.

References

1. Abdolshah, M.: A review of resource-constrained project scheduling problems (RCPSP) approaches and solutions. Int. Trans. J. Eng. Manag. Appl. Sci. Technol. **5**(4), 253–286 (2014)
2. Alcaraz, J., Maroto, C.: A hybrid genetic algorithm based on intelligent encoding for project scheduling. In: Józefowska, J., Weglarz, J. (eds.) Perspectives in Modern Project Scheduling, pp. 249–274. Springer, Boston (2006). https://doi.org/10.1007/978-0-387-33768-5_10
3. Berthaut, F., Pellerin, R., Hajji, A., Perrier, N.: A path relinking-based scatter search for the resource-constrained project scheduling problem. Int. J. Project Organ. Manag. **10**(1), 1–36 (2018)
4. Blazewicz, J., Lenstra, J.K., Kan, A.R.: Scheduling subject to resource constraints: classification and complexity. Discrete Appl. Math. **5**(1), 11–24 (1983)
5. Brucker, P., Drexl, A., Möhring, R., et al.: Resource-constrained project scheduling: notation, classification, models, and methods. Eur. J. Oper. Res. **112**(1), 3–41 (1999)
6. Carlier, J., Moukrim, A., Xu, H.: A memetic algorithm for the resource constrained project scheduling problem. In: Proceedings of the International Conference on Industrial Engineering and Systems Management, IESM (2009)
7. Chen, D., Liu, S., Qin, S.: Memetic algorithm for the resource-constrained project scheduling problem. In: Proceeding of the 11th World Congress on Intelligent Control and Automation, WCICA, pp. 4991–4996. IEEE (2014)
8. Czogalla, J., Fink, A.: Particle swarm topologies for resource constrained project scheduling. In: Krasnogor, N., et al. (eds.) NICSO 2008, pp. 61–73. Springer-Verlag, Berlin Heidelberg (2009). https://doi.org/10.1007/978-3-642-03211-0_6
9. Debels, D., De Reyck Leus, B.R., Vanhoucke, M.: A hybrid scatter search electromagnetism meta-heuristic for project scheduling. Eur. J. Oper. Res. **169**, 638–653 (2006)
10. Debels, D., Vanhoucke, M.: Decomposition-based genetic algorithm for the resource-consrtained project scheduling problem. Oper. Res. **55**, 457–469 (2007)
11. Elsayed, S., Sarker, R., Ray, T., Coello, C.C.: Consolidated optimization algorithm for resource-constrained project scheduling problems. Inf. Sci. **418–419**, 346–362 (2017)
12. Fang, C., Wang, L.: An effective shuffled frog-leaping algorithm for resource-constrained project scheduling problem. Comput. Oper. Res. **39**(5), 890–901 (2012)

13. Gagnon, M., Boctor, F.F., d'Avignon, G.: A tabu search algorithm for the resource-constrained project scheduling problem. In: ASAC (2004)
14. Gimadi, E.K.: On some mathematical models and methods for planning large-scale projects. Models and Optimization Methods. In: Proceedings an USSR Sib. Branch, Math. Inst., Novosibirsk. Nauka, vol. 10, pp. 89–115 (1988)
15. Gimadi, E.K., Goncharov, E.N., Mishin, D.V.: On some implementations of solving the resource-constrained project scheduling problem. Yugoslav J. Oper. Res. **29**(1), 31–42 (2019)
16. Gimadi, E.K., Zalyubovskii, V.V., Sevast'yanov, S.V.: Polynomial solvability of scheduling problems with storable resources and deadlines. Diskretnyi Analiz i Issledovanie Operazii **7**(1), 9–34 (2000)
17. Goncalves, J., Resende, M.G.C., Mendes, J.: A biased random key genetic algorithm with forward-backward improvement for resource-constrained project scheduling problem. J. Heuristics. **17**, 467–486 (2011). https://doi.org/10.1007/s10732-010-9142-2
18. Goncharov, E.N., Leonov, V.V.: Genetic algorithm for the resource-constrained project scheduling problem. Autom Remote Control **78**(6), 1101–1114 (2017). https://doi.org/10.1134/S0005117917060108
19. Goncharov, E.N.: Variable neighborhood search for the resource constrained project scheduling problem. In: Bykadorov, I., Strusevich, V., Tchemisova, T. (eds.) MOTOR 2019. CCIS, vol. 1090, pp. 39–50. Springer, Cham (2019). https://doi.org/10.1007/978-3-030-33394-2_4
20. Goncharov, E.N.: Local search algorithm for the resource-constrained project scheduling problem. Diskret. Anal. Issled. Oper. **29**(4), 15–37 (2022)
21. Hartmann, S., Briskorn, D.: A survey of variants and extentions of the resource-constrained project scheduling problem. Eur. J. Oper. Res. **207**, 1–14 (2010)
22. Herroelen, W., Demeulemeester, E., De Reyck, B.: A Classification Scheme for Project Scheduling. In: Weglarz J. (Ed.). Project Scheduling-Recent Models, Algorithms and Applications, International Series in Operations Research and Management Science. Kluwer Academic Publishers, Dordrecht, vol. 14, no. 1, pp. 77–106 (1998)
23. Ismail, I.Y., Barghash, M.A.: Diversity guided genetic algorithm to solve the resource constrained project scheduling problem. Int. J. Plan. Sched. **1**(3), 147–170 (2012)
24. Jedrzejowicz, P., Ratajczak, E.: Population learning algorithm for the resource-constrained project scheduling. In: Józefowska, J., Weglarz, J. (eds.) Perspectives in Modern Project Scheduling, pp. 275–296. Springer, Boston (2006). https://doi.org/10.1007/978-0-387-33768-5_11
25. Kochetov, Y., Stolyar, A.: Evolutionary local search with variable neighborhood for the resource-constrained project scheduling problem. In: Proceedings of the 3th International Workshop of Computer Science and Information Technologies, vol. 96–99 (2003)
26. Kolisch, R., Hartmann, S.: Experimental investigation of heuristics for resource-constrained project scheduling: an update. Eur. J. Oper. Res. **174**, 23–37 (2006)
27. Kolisch, R., Hartmann, S.: Heuristic Algorithms for Solving the Resource-Constrained Project Scheduling Problem: Classification and Computational Analysis. In: Weglarz J., (ed). Project scheduling: Recent models, Algorithms and Applications. Kluwer Academic Publishers, pp. 147–178 (1999)
28. Kolisch, R., Sprecher, A., Drexl, A.: Characterization and generation of a general class of resource-constrained project scheduling problems. Manag. Sci. **41**, 1693–1703 (1995)

29. Kolisch, R., Sprecher, A.: PSPLIB – a project scheduling problem library. Eur. J. Oper. Res. **96**, 205–216 (1996). http://www.om-db.wi.tum.de/psplib/

30. Koulinas, G., Kotsikas, L., Anagnostopoulos, K.: A particle swarm optimization based hyper-heuristic algorithm for the classic resource constrained project scheduling problem. Inf. Sci. **277**, 680–693 (2014)

31. Lim, A., Ma, H., Rodrigues, B., Tan, S.T., Xiao, F.: New meta-heuristics for the resource-constrained project scheduling problem. Flex. Serv. Manuf. J. **25**(1–2), 48–73 (2013). https://doi.org/10.1007/s10696-011-9133-0

32. Mendes, J.J.M., Goncalves, J.F., Resende, M.G.C.: A random key based genetic algorithm for the resource constrained project scheduling problem. Comput. Oper. Res. **36**, 92–109 (2009)

33. Mobini, M.D.M., Rabbani, M., Amalnik, M.S., et al.: Using an enhanced scatter search algorithm for a resource-constrained project scheduling problem. Soft Comput. **13**, 597–610 (2009). https://doi.org/10.1007/s00500-008-0337-5

34. Mobini, M., Mobini, Z., Rabbani, M.: An artificial immune algorithm for the project scheduling problem under resource constraints. Appl. Soft Comput. **11**(2), 1975–1982 (2011)

35. Palpant, M., Artigues, C., Michelon, P.: Solving the resource-constrained project scheduling problem with large neighborhood search. Ann. Oper. Res. **131**, 237–257 (2004)

36. Pellerin, R., Perrier, N., Berthaut, F.: LSSPER: A survey of hybrid metaheuristics for the resource-constrained project scheduling problem. Eur. J. Oper. Res. **280**(2), 395–416 (2020)

37. Proon, S., Jin, M.: A genetic algorithm with neighborhood search for the resource-consrtained project scheduling problem. Naval Res. Logist. **58**, 73–82 (2011)

38. Ranjbar, M., Kianfar, F.: A hybrid scatter search for the RCPSP. Sci. Iranica **16**(1), 11–18 (2009)

39. Paraskevopoulos, D.C., Tarantilis, C.D., Ioannou, G.: Solving project scheduling problems with resource constraints via an event list-based evolutionary algorithm. Expert Syst. Appl. **39**(4), 3983–3994 (2012)

40. Valls, V., Ballestin, F., Quintanilla, M.S.: Justification and RCPSP: a technique that Pays. Eur. J. Oper. Res. **165**, 375–386 (2005)

41. Valls, V., Ballestin, F., Quintanilla, S.: A hybrid genetic algorithm for the resource-consrtained project scheduling problem. Eur. J. Oper. Res. **185**(2), 495–508 (2008)

42. Valls, V., Ballestin, F., Quintanilla, S.: A population-based approach to the resource-constrained project scheduling problem. Ann. Oper. Res. **131**, 305–324 (2004). https://doi.org/10.1023/B:ANOR.0000039524.09792.c9

43. Vanhoucke, M.: Resource-constrained project scheduling. In: Project Management with Dynamic Scheduling. Springer-Verlag, Heidelberg, pp. 107–137 (2012). https://doi.org/10.1007/978-3-642-25175-7_7

44. Chen, W., Shi, Y.J., Teng, H.F., et al.: An efficient hybrid algorithm for resource-constrained project scheduling. Inf. Sci. **180**(6), 1031–1039 (2010)

45. Wang, H., Li, T., Lin, T.: Efficient genetic algorithm for resource-constrained project scheduling problem. Trans. Tianjin Univ. **16**(5), 376–382 (2010). https://doi.org/10.1007/s12209-010-1495-y

46. Weglarz, J.: Project Scheduling: Recent Models, Algorithms and Applications. Kluwer Academic Publishers, Boston (1999)

47. Ying, K.C., Lin, S.W., Lee, Z.J.: Hybrid-directional planning: improving improvement heuristics for scheduling resource-constrained projects. Int. J. Adv. Manuf. Technol. **41**(3–4), 358–366 (2009). https://doi.org/10.1007/s00170-008-1486-5

48. Zamani, R.: A competitive magnet-based genetic algorithm for solving the resource-constrained project scheduling problem. Eur. J. Oper. Res. **229**(2), 552–559 (2013)
49. Zheng, X., Wang, L.: A multi-agent optimization algorithm for resource constrained project scheduling problem. Expert Syst. Appl. **42**(15–16), 6039–6049 (2015)

Nonlocal Optimization Methods for Nonlinear Controlled Systems with Terminal Constraints

Dmitry Trunin[(✉)] [iD]

Buryat State University, Ulan-Ude, Russia
tdobsu@yandex.ru

Abstract. A new approach to optimization of nonlinear control systems with terminal constraints based on the sequential solution of nonlocal control improvement problems in the form of special systems of functional equations in the control space is considered. The corresponding systems are constructed as fixed point problems of special control operators with an additional algebraic equation, to the solution of which the apparatus of the theory and methods of fixed points is applied. The proposed algorithms for successive approximations of control with the preservation of all terminal constraints at each iteration of approximations do not contain the laborious operation of parametric variation to improve control, which is typical for gradient improvement methods. The effectiveness of the proposed methods for constructing relaxation control sequences on the set of admissible controls in the considered class of optimization of control systems is illustrated by model examples.

Keywords: Nonlinear controlled system · Terminal constraints · Conditions of control improvement · Fixed point problem · Iterative algorithm

1 Introduction

To solve nonlinear optimal control problems, iterative methods of successive improvement of admissible controls based on the necessary optimality conditions (maximum principle, differential maximum principle, including in projection form, etc.) are traditionally applied. Characteristic representatives of the considered class of methods are well-known gradient methods—methods of conditional gradient and gradient projection.

The most laborious in standard gradient methods of optimal control is the procedure of varying the control in a small neighborhood of the current approximation at each iteration to improve the current approximation of the control.

The paper [1] proposes methods for nonlocal improvement of controls in the class of linear in state optimal control problems with a free right end with a linear and quadratic in state objective functional. These methods are based on special formulas for the increment of the objective functional without residual expansion terms and do not contain the time-consuming operation of parametric variation of the control in the vicinity of the current approximation. Improvement of control is achieved at the cost of solving two special Cauchy problems. The indicated features of the methods are essential factors for increasing the efficiency of solving problems of the class under consideration.

In [2], methods for nonlocal improvement of control were developed in the class of optimal control problems polynomial in state with a free right end, generalizing the methods of [1]. These methods are based on formulas for the increment of the objective functional without residual terms of the increments, which were obtained with the help of modifications of the conjugate system. In this case, to improve the control, it is required to solve a special boundary value improvement problem. To solve this boundary value problem, the well-known perturbation approach in mathematics is used.

In the paper [3], the methods of nonlocal improvement [2] are generalized for a class of optimal control problems that are non-linear in state with a free right end. In the indicated class of problems, the modified conjugate system is a differential-algebraic system (the Cauchy problem with additional algebraic relations).

In this paper, the methods of [3] are generalized to a class of optimal control problems, nonlinear in state, with terminal constraints. For non-local improvement of admissible controls in the considered class of problems, it is proposed to use an iterative method for solving a system of functional equations in the control space that determines the conditions for non-local improvement, which is considered as a special fixed point problem.

At present, the development of effective methods for solving optimal control problems with constraints that arise in the modeling of natural science processes and in other applications is one of the topical mathematical problems of control theory, which is the subject of numerous works, in particular [4–6].

2   Control Improvement Problem

We consider a class of optimal control problems that are nonlinear in state and linear in control with one terminal equality constraint

$$\dot{x} = f(x, u, t), t \in T = [t_0, \ t_1], \tag{1}$$

$$x(t_0) = x^0, \tag{2}$$

$$u(t) \in U \subset R^r, t \in T, \tag{3}$$

$$\Phi_0(u) = \varphi(x(t_1)) + \int_T F(x, u, t)dt \to \min, \tag{4}$$

$$\Phi_1(u) = \chi(x(t_1)) = 0. \tag{5}$$

In problem (1)–(5):

$x = (x_1(t), x_2(t), \ldots x_n(t))$ is a state vector,

$u = (u_1(t), u_2(t), \ldots u_r(t))$ is a control vector.

Initial state $x^0 \in R^n$ is given.

The functions $f(x, u, t)$ and $F(x, u, t)$ are nonlinear in x and linear in u

$$f(x, u, t) = A(x, t)u + b(x, t),$$

$$F(x, u, t) = \langle d(x, t), u \rangle + g(x, t).$$

The functions $A(x, t)$, $b(x, t)$, $d(x, t)$ and $g(x, t)$ are nonlinear and differentiable in x and continuous in u on the set $R^n \times T$; the functions $\varphi(x)$ and $\chi(x)$ are nonlinear and differentiable in x; U is a compact and convex set; time interval T is fixed.

Various optimal control problems with terminal, phase and mixed constraints can be reduced to the form (1)–(5).

By accessible controls in problem (1)–(5) we mean functions that are piecewise continuous on an interval T and have values in a compact and convex set $U \subset R^r$:

$$V = \{u \in PC^r(T) : u(t) \in U, t \in T\}.$$

For accessible control $v \in V$ we denote $x(t, v), t \in T$ the solution of the Cauchy problem (1), (2) for $u = v(t), t \in T$.

By admissible controls W we mean accessible controls if the terminal constraint is satisfied (5):

$$W = \{u \in V : \chi(x(t_1, u)) = 0\}.$$

In problem (1)–(5) the Pontryagin function with the conjugate variable $p \in R^n$ can be presented in the form:

$$H(p, x, u, t) = H_0(p, x, t) + \langle H_1(p, x, t), u \rangle,$$

where $H_0(p, x, t) = \langle p, b(x, t) \rangle - g(x, t)$, $H_1(p, x, t) = A(x, t)^T p - d(x, t)$.

Consider the regular Lagrange functional:

$$L(u, \lambda) = \Phi_0(u) + \lambda \Phi_1(u), \lambda \in R.$$

Following [3], the formula for the increment of the Lagrange functional, which does not contain the remainder of the expansion, takes the form:

$$\Delta_v L(u^0, \lambda) = -\int_T \langle H_1(p(t, u^0, v, \lambda), x(t, v), t), v(t) - u^0(t) \rangle dt, \qquad (6)$$

where (u^0, v) are accessible controls; $p(t, u^0, v, \lambda), t \in T$ the solution of the modified differential-algebraic conjugate system

$$\dot{p} = -H_x(p, x, u, t) - r(t), \qquad (7)$$

$$\langle H_x(p, x(t, u^0), u^0(t), t), x(t, v) - x(t, u^0) \rangle + \langle r(t), x(t, v) - x(t, u^0) \rangle = \\ = H(p, x(t, v), u^0(t), t) - H(p, x(t, u^0), u^0(t), t), \qquad (8)$$

$$p(t_1) = -\varphi_x(x(t_1, u^0)) - \lambda \chi_x(x(t_1, u^0)) - q, \qquad (9)$$

$$\langle \varphi_x(x(t_1, u^0)) + \lambda \chi_x(x(t_1, u^0)), x(t_1, v) - x(t_1, u^0) \rangle +$$
$$+ \langle q, x(t_1, v) - x(t_1, u^0) \rangle = \quad (10)$$
$$= \varphi(x(t_1, v)) - \varphi(x(t_1, u^0)) + \lambda (\chi(x(t_1, v)) - \chi(x(t_1, u^0))).$$

Algebraic relations (8), (10) can always be solved in analogy with [3] with respect to the values $r(t)$, q and reduce the differential-algebraic problem to an ordinary differential problem (perhaps in a non-unique way).

In particular, in the subclass of problems quadratic in state (the functions f, F, φ, χ are quadratic in x) the values $r(t)$, q can be represented by the following relations:

$$r(t) = \frac{1}{2} H_{xx}(p, x(t, u^0), u^0(t), t)(x(t, v) - x(t, u^0)),$$

$$q = \frac{1}{2}(\varphi_{xx}(x(t_1, u^0)) + \lambda \chi_{xx}(x(t_1, u^0)))(x(t_1, v) - x(t_1, u^0)).$$

For accessible control $u^0 \in V$ and a fixed projection parameter $\alpha > 0$ similarly to [2], we form the vector function

$$u^\alpha(p, x, t) = P_U\left(u^0(t) + \alpha H_1(p, x, t)\right), \ p \in R^n, \ x \in R^n, \ t \in T, \ \alpha > 0,$$

where P_U is an operator of projection onto a set U in the Euclidean norm.

According to the well-known property of the projection, the following estimate holds:

$$\int_T \langle H_1(p, x, t), u^\alpha(p, x, t) - u^0(t) \rangle dt \geq \frac{1}{\alpha} \int_T ||u^\alpha(p, x, t) - u^0(t)||^2 dt. \quad (11)$$

Then from (6) and (11) follows the estimate of the increment of the functional:

$$\Delta_v L(u^0, \lambda) \leq -\frac{1}{\alpha} \int_T ||u^\alpha(p, x, t) - u^0(t)||^2 dt. \quad (12)$$

We pose the problem of improving the admissible control $u^0 \in W$: find the control $v \in W$ with property

$$\Phi_0(v) \leq \Phi_0(u^0).$$

Let us show that for a nonlocal improvement of the admissible control $u^0 \in W$ it suffices to solve for some $\alpha > 0$ the following system of functional equations in the control space:

$$v(t) = u^\alpha(p(t, u^0, v, \lambda), x(t, v), t), t \in T, \lambda \in R,$$
$$\chi(x(t_1, v)) = 0. \quad (13)$$

Let the control v is a solution of the system (13). It is easy to see that $v \in W$. Then, by virtue of estimate (12), there is an improvement in the objective functional Φ_0 with an estimate

$$\Delta_v \Phi_0(u^0) \leq -\frac{1}{\alpha} \int_T ||v(t) - u^0(t)||^2 dt. \quad (14)$$

It follows from the estimate (14) that if control v differs from control u^0, then a strict improvement of the target functional is provided.

The system of equations (13) is considered as a fixed point problem in the control space with an additional algebraic equation. This allows us to apply and modify the well-known iterative fixed-point methods to solve the system (13).

3   Iterative Methods

To solve system (13) for fixed $\alpha > 0$, the following modification of the well-known algorithm of the simple iteration method [7] is proposed for $k \geq 0$:

$$v^{k+1}(t) = u^\alpha(p(t, u^0, v^k, \lambda^k), x(t, v^{k+1}), t), t \in T, \lambda \in R,$$
$$\chi(x(t_1, v^{k+1})) = 0. \tag{15}$$

As an initial approximation of the iterative process (15), the control $v^0 \in V$ is chosen. The main feature of the proposed iterative algorithm is the selection of the parameter $\lambda \in R$ at each iteration for $k \geq 1$ to satisfy the terminal constraint. It is assumed that such a possibility exists.

The implementation of the proposed implicit iterative process (15) at each iteration consists in the following actions.

Find the solution $p^\lambda(t)$, $t \in T$ of the problem (7)–(10) for $v = v^k(t)$.

Let $x^\lambda(t)$, $t \in T$ the solution to the special Cauchy problem:

$$\dot{x} = f(x, u^\alpha(p^\lambda(t), x, t), t),\ t \in T,\ x(t_0) = x^0.$$

Find the value of the Lagrange multiplier $\overline{\lambda} \in R$ from the condition:

$$\chi(x^\lambda(t_1)) = 0. \tag{16}$$

We form the next control approximation according to the rule:

$$v^{k+1}(t) = u^\alpha(p^{\overline{\lambda}}(t), x^{\overline{\lambda}}(t), t), t \in T.$$

Thus, the implementation of the implicit process (15) at each iteration is reduced to solving the algebraic equation (16).

Another modification of the algorithm of the simple iteration method for solving system (13) has a more familiar standard explicit form for $k \geq 0$:

$$v^{k+1}(t) = u^\alpha(p(t, u^0, v^k, \lambda^k), x(t, v^k), t), t \in T, \lambda \in R,$$
$$\chi(x(t_1, v^{k+1})) = 0. \tag{17}$$

For this modification, at each iteration of the process (17), after calculating the solution $p^\lambda(t)$, $t \in T$ of the problem (7)–(10) for $v = v^k(t)$ auxiliary control is formed

$$v^\lambda(t) = u^\alpha(p^\lambda(t), x(t, v^k), t), t \in T.$$

For the auxiliary control v^λ a solution $x(t, v^\lambda)$, $t \in T$ to the standard Cauchy problem is found

$$\dot{x} = f(x, v^\lambda(t), t),\ t \in T,\ x(t_0) = x^0.$$

The value of the Lagrange multiplier $\lambda \in R$ at each iteration of the process (17) is selected from the condition of fulfillment of the terminal constraint:

$$\chi(x(t_1, v^\lambda)) = 0. \tag{18}$$

For the obtained solution $\overline{\lambda} \in R$ of equation (18) the following control approximation is determined

$$v^{k+1}(t) = v^{\overline{\lambda}}(t), t \in T.$$

A feature of the proposed iterative algorithms for solving the fixed point problem (13) is the fulfillment of the terminal constraint (5) at each iteration of the process of successive approximations of the control. In this case, the initial approximation v^0 of iterative processes may not satisfy the terminal constraint, which is important for the practical implementation of algorithms.

The convergence of the proposed iterative processes is regulated by the choice of the projecting parameter $\alpha > 0$ and can be substantiated on the basis of the perturbation method and the contraction mapping principle similarly to [2] for sufficiently small values of $\alpha > 0$.

Iterative processes are applied until the first improvement in control u^0. Next, a new improvement task is constructed for the obtained control, and the process is repeated. The criterion for stopping the control improvement iterations is the absence of strict control improvement in terms of the target functional.

Based on the sequential solution of control improvement problems, the corresponding iterative methods for constructing relaxation control sequences that satisfy the terminal constraint are formed.

4   Example

This section presents the results of calculations of the model problem of satellite rotation stabilization [8] using the proposed method of nonlocal improvement (13) based on the implicit iterative process (15). A comparative analysis of the effectiveness of this method (M3) with the standard methods of conditional gradient (M1) and gradient projection (M2) is carried out [9].

The parameter $\alpha > 0$, which regulates the convergence of the proposed method (M3), was chosen experimentally according to the rule of change by one order, starting from the value $\alpha = 1$.

The calculation of phase and conjugate Cauchy problems was carried out using the standard Fortran procedure *divprk* [10], which implements the 5th-6th order Runge-Kutta method. The absolute error of the numerical integration of the Cauchy problems was set equal to 10^{-10}. The values of the calculated control, phase and conjugate variables during the calculation were stored in the nodes of a given uniform grid with a discretization step equal to 0.001. Piecewise constant interpolation was used to approximate the controls between the nodes of the uniform grid.

The solution of the algebraic equation with respect to the Lagrange multiplier arising at each iteration of the process (15) was carried out by minimizing the quadratic residual of the equation by means of the standard Fortran procedure *dumpol* [10], which implements the deformable polyhedron method.

The laboriousness of the methods was estimated by the total number of Cauchy calculation problems for phase and conjugate variables.

For a numerical solution, the problem [8] was reduced to the form (1)–(5):

$$\dot{x}_1 = \frac{1}{3}x_2 x_3 + 100 u_1,$$

$$\dot{x}_2 = -x_1 x_3 + 25 u_2,$$

$$\dot{x}_3 = -x_1 x_2 + 100 u_3,$$

$$x_1(0) = 200, \ x_2(0) = 30, \ x_3(0) = 40,$$

$$u_1(t) \in [-40, \ 40], \ u_2(t) \in [-20, \ 20], u_3(t) \in [-40, \ 40], \ t \in [0, \ 0.1],$$

$$\Phi_0(u) = \frac{1}{2}\left(x_2^2(0.1) + x_3^2(0.1)\right) \to \min,$$

$$\Phi_1(u) = x_1(0.1) = 0.$$

The equations of the system describe the dynamics of the rotation of a satellite equipped with three jet engines. Controls characterize fuel consumption. The minimized functional from the control reflects the goal to achieve a state characterized by the absence of satellite rotation (stabilization).

As an initial approximation for all compared methods, we chose $u(t) \equiv 0, t \in T$.

The practical criterion for stopping the calculation of the problem in all methods was the condition

$$|\Phi_0(u^{k+1}) - \Phi_0(u^k)| \le M|\Phi_0(u^k)|,$$

where $k > 0$ is an iteration number, $M = 10^{-5}$.

Comparative qualitative and quantitative results of calculations are presented in Table 1.

Table 1. Results of calculations by compared methods.

Method	Φ_0^*	Φ_1^*	N	Note
M1	3.16428×10^{-13}	2.45074×10^{-7}	8512	0.5
M2	1.48471×10^{-13}	3.13041×10^{-7}	2642	0.5
M3	3.63122×10^{-13}	5.229144×10^{-8}	1458	10^{-5}

In Table 1 Φ_0^* is a calculated value of the objective functional of the problem, Φ_1^* is a module of the calculated value of the functional-constraint, N is a total number of solved phase and conjugate Cauchy problems. The note for methods M1 and M2 indicates the value of the penalty parameter, for method

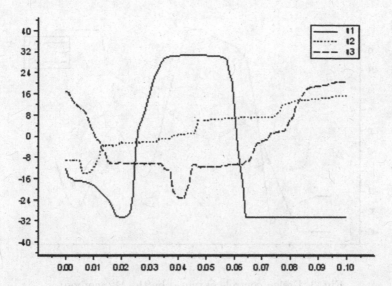

Fig. 1. Design controls obtained by the M1 method.

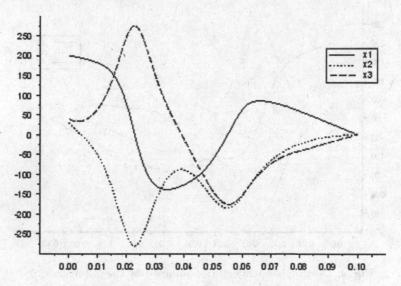

Fig. 2. Design phase trajectories obtained by the M1 method.

M3 the value of the projection parameter α, which ensures the convergence of the iterative process (15).

Graphs of computational controls and phase trajectories are given respectively on Fig. 1, 2 (M1), Fig. 3, 4 (M2), Fig. 5, 6 (M3).

The calculation data allow us to conclude that the nonlocal method (M3) has better computational efficiency, estimated by the total number of Cauchy problems, compared to the conditional gradient and gradient projection meth-

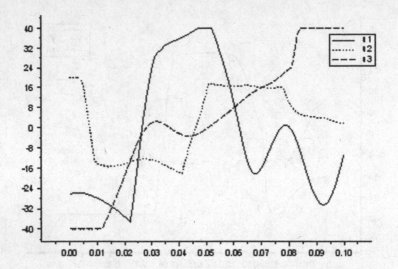

Fig. 3. Design controls obtained by the M2 method.

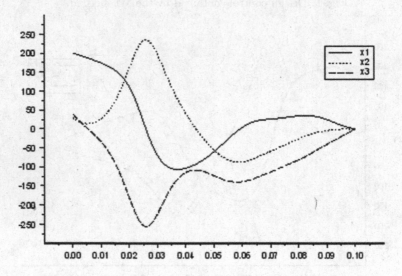

Fig. 4. Design phase trajectories obtained by the M2 method.

ods. In this case, almost identical calculated phase trajectories and values of the control functionals were obtained.

In addition, the proposed method (M3) achieves a terminal constraint with a given accuracy that is better than the conditional gradient and gradient projection methods.

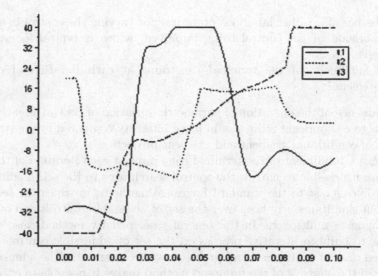

Fig. 5. Design controls obtained by the M3 method.

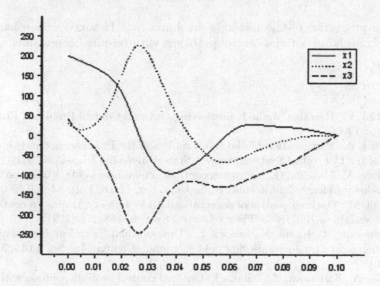

Fig. 6. Design phase trajectories obtained by the M3 method.

5 Conclusion

The proposed methods for nonlocal improvement of admissible controls in the considered class of nonlinear problems with constraints are characterized by the following properties:

1. the absence of a rather laborious procedure for varying the control in a small neighborhood of the control being improved, which is typical for gradient methods;
2. exact fulfillment of the terminal constraint at each iteration of control improvement.

The absence of the operation of parametric variation of control at each iteration leads to a significant reduction in labor intensity compared to the standard methods of conditional gradient and gradient projection.

The exact fulfillment of the terminal constraint at each iteration of the process makes it possible to narrow the control search area to the set of admissible controls, in contrast to the standard Lagrange methods, in which the search is carried out simultaneously both over the set of accessible controls and over the set of Lagrange multipliers. In the general case, penalty methods also do not allow one to build an iterative process on the set of admissible controls in the considered class of optimal control problems with constraints. The admissibility property of the controls of the proposed method makes it possible to effectively obtain controls that are acceptable in practice according to the optimality criterion.

These properties of the methods are important factors for increasing the efficiency of solving optimal control problems with terminal constraints.

References

1. Srochko, V.: Iterative Methods for Solving Optimal Control Problems. Fizmatlit, Moscow (2000)
2. Buldaev, A.: Perturbation Methods in Problem of the Improvement and Optimization of the Controlled Systems. Buryat State University, Ulan-Ude (2008)
3. Buldaev, A., Morzhin, O.: Improving controls in nonlinear systems based on boundary value problems. Bull. Irkutsk State Univ. Ser. Math. 1(2), 94–106 (2009)
4. Levskii, M.: Optimal control of pacecraft attitude with constraints on control and phase variables. Bull. RAS Theory Control Syst. 6, 158–176 (2021)
5. Chertovskih, R., Khalil, N., Pereira, F.: Time-optimal control problem with state constraints in a time-periodic flow field. Commun. Comput. Inf. Sci. 1145, 340–354 (2020)
6. Diveev, A., Sofronova, E., Zelinka, I.: Optimal control problem solution with phase constraints for group of robots by Pontryagin maximum principle and evolutionary algorithm. Mathematics 8, 2105 (2020)
7. Samarskii, A., Gulin, A.: Numerical Methods. Nauka, Moscow (1989)
8. Tyatushkin, A.: Numerical Methods and Software Tools for Optimizing Controlled Systems. Nauka, Novosibirsk (1992)
9. Vasiliev, O.: Optimization Methods. World Federation Publishers Company Inc., Atlanta (1996)
10. Bartenev, O.: Fortran for Professionals. IMSL Math Library. Part 2. Dialog-MIFI, Moscow (2001)

Discrete and Combinatorial Optimization

Three-Bar Charts Packing Problem

Adil Erzin[1,2,3]($\boxtimes$) [iD] and Konstantin Sharankhaev[2]

[1] Sobolev Institute of Mathematics, SB RAS, Novosibirsk 630090, Russia
[2] Novosibirsk State University, Novosibirsk 630090, Russia
`k.sharankhaev@g.nsu.ru`
[3] St. Petersburg State University, St. Petersburg 199034, Russia
`adilerzin@math.nsc.ru`

Abstract. Three-Bar Charts Packing Problem is to pack the bar charts consisting of three bars each into the horizontal unit-height strip of minimal length. The bars of each bar chart may move vertically within the strip, but it is forbidden to change the order and separate the bars. For this novel issue, which is a generalization of the strongly NP-hard Two-Bar Charts Packing Problem considered earlier, we propose several approximation algorithms with guaranteed accuracy.

Keywords: Bar charts · Strip packing · Approximation

1 Introduction

To avoid discrepancies, hereinafter we will understand the *length* as a horizontal, and the *height* as a vertical size. The problem of packing bar charts in a strip was first described in [8] and can be formulated as follows. Let us have a set of bar charts (BCs) consisting of several unit-length bars. The height of each bar is positive, but does not exceed 1. In the Bar Charts Packing Problem (BCPP), it is required to pack all BCs in a unit-height horizontal strip of minimal length. Let us split the strip into the equal unit-length unit-height rectangles – the "cells" and number them by positive integers $1, 2, \ldots$ Then the packing length is the number of cells with at least one bar. In a *feasible* packing, the bars of each BC do not change order and must occupy the adjacent cells, but they can move vertically independently of each other within the strip. Moreover, in each cell, the total height of the bars in it should not exceed 1. Further, we will consider only feasible packings and will omit the word "feasible".

Let us denote the BCPP of packing BCs consisting of k bars each (k-BCs) as k-BCPP. Earlier in [9–11], a 2-BCPP was considered, which is a generalization of the Bin Packing Problem (BPP) [14] and 2-Dimensional Vector Packing Problem [16]. Several approximation algorithms were proposed with a priori guaranteed estimates. In [9] was proposed an $O(n)$-time algorithm which constructs a packing of n arbitrary 2-BCs of length at most $2 \cdot OPT + 1$, where OPT

The research was supported by the Russian Science Foundation (grant No. 22-71-10063 "Development of intelligent tools for optimization multimodal flow assignment systems in congested networks of heterogeneous products").

is an optimum of the 2-BCPP (Later, the additive constant was removed and the estimate was reduced to $2 \cdot OPT$). When at least one bart of each 2-BC is higher than $1/2$ ("big" 2-BCs), an $O(n^3)$-time, $3/2$-approximation algorithm was proposed. If all 2-BCs are big and non-increasing or non-decreasing, the complexity was reduced to $O(n^{2.5})$ preserving the ratio [10]. Paper [11] updates the estimates for the packing length of big 2-BCs, keeping the time complexity. In [11] a $5/4$-approximation $O(n^{2.5})$-time algorithm for packing big non-increasing or non-decreasing 2-BCs was presented. For the case of big 2-BCs (not necessarily non-increasing or non-decreasing), a $16/11$-approximation $O(n^3)$-time algorithm was proposed.

This paper considers a 3-BCPP of packing BCs consisting of 3 bars each (3-BCs). Let us introduce the following definitions.

Definition 1. Packing *is a function* $p : S \to Z^+$, *which associates with each 3-BC i the cell number of the strip $p(i)$ into which the first bar of 3-BC i falls.*

As a result of packing p, the bars of 3-BC i occupy the cells $p(i)$, $p(i) + 1$ and $p(i) + 2$.

Definition 2. *The packing length $L(p)$ is the number of strip cells in which at least one bar falls.*

Definition 3. *Two BCs form a t-union if t cells of the strip contain the bars of both bar charts.*

Two 3-BCs can form 0-, 1-, 2- and 3-unions.

The 3-BCPP generalizes the BPP. If all bars of each 3-BC are equal, then 3-BCPP is the same as BPP, which is strongly NP-hard and inapproximable within the $3/2 - \varepsilon$, for any $\varepsilon > 0$ [19]. However, for BPP, several approximation algorithms are known [2,7,14,15,17,20]. The best-known estimate on the number of bins used is $71/60 \cdot OPT + 1$, where OPT is optimum of BPP [21]. The known results for BPP can be used to get an approximate packing for 3-BCPP; it is sufficient to put each 3-BC in a minimal rectangle. However, the resulting estimate is rough.

3-Dimensional Vector Packing Problem (3-DVPP) is a generalization of BPP and a particular case of 3-BCPP when only 3-unions are allowed. It considers three attributes for each item and bin. The problem is to pack all items in the minimum number of bins, considering three attributes of bin's capacity limits [3,6,16]. Applying an approximation algorithms to the 3-DVPP gives a feasible packing for 3-BCPP, but it is inaccurate.

1.1 Our Contribution

This paper presents a formulation of 3-BCPP. We find a non-trivial accuracy estimates for the general case and particular cases of 3-BCPP. To formulate our results, we need the following definition.

Definition 4. *If at least k bars of the BC are higher than $1/2$, then we call such BC "k-big". If BC is k-big, then it is also $(k - 1)$-big.*

(i) We propose an $O(n)$-time algorithm A_3 to construct a packing of length at most $3 \cdot OPT + 2$, where OPT is the optimum of the 3-BCPP.

(ii) If all 3-BCs are 2-big, the algorithm M_w [10] yields a 5/4-approximate solution with time complexity $O(n^3)$. If additionally, in each 2-big 3-BC a middle bar is big, we show how to find a 9/8-approximate packing with time complexity $O(n^{2,5})$.

(iii) If all 3-BCs are 1-big, we prove that the 3-BCPP remains strongly NP-hard. The complexity of packing 2-big 3-BCs we do not know yet.

The rest of the paper is organized as follows. Section 2 provides a statement of the 3-BCPP as Boolean Linear Programming. Section 3 contains a proof of the NP-hardness of packing 1-big 3-BCs. In Sect. 4, we describe the algorithms under consideration. Section 5 contains new approximation results, and the last section concludes the paper.

2 Formulation of the Problem

Let us have a semi-infinite unit-height horizontal strip and a set of three-bars charts (3-BCs) S, $|S| = n$. Each 3-BC $i \in S$, consists of three unit-length bars. The height of the first bar is $a_i \in (0,1]$, of the second is $b_i \in (0,1]$ and of the third is $c_i \in (0,1]$. Let us split the strip into identical rectangles of unit length and height, which we call the "cells", starting from the beginning of the strip, and number them with positive integers $1, 2, \ldots$

We introduce the following variables.

$$x_{ij} = \begin{cases} 1, \text{ if the first bar of 3-BC } i \text{ is in the cell } j; \\ 0, \text{ else.} \end{cases}$$

$$y_j = \begin{cases} 1, \text{ if the cell } j \text{ contains at least one bar;} \\ 0, \text{ else.} \end{cases}$$

Then 3-BCPP is as follows.

$$\sum_j y_j \rightarrow \min_{x_{ij}, y_j \in \{0,1\}}; \tag{1}$$

$$\sum_j x_{ij} = 1, \ i \in S; \tag{2}$$

$$\sum_i a_i x_{ij} + \sum_k b_k x_{kj-1} + \sum_l c_l x_{lj-2} \leq y_j, \ \forall j. \tag{3}$$

In this formulation, criterion (1) is the minimization of the packing length. Constraints (2) require each 3-BC to be packed into a strip once. Constraints (3) ensure that the sum of the bar's heights in any cell does not exceed 1, and also link two groups of variables.

The 3-BCPP (1)–(3) is strongly NP-hard as a generalizations of the BPP [14]. Moreover, the problem is $(3/2 - \varepsilon)$-inapproximable for any $\varepsilon > 0$ unless P=NP [19].

3   NP-Hardness of Packing 1-Big 3-BCs

Since the complexity of the problem in the particular case when all 3-BCs are 1-big was not known, we prove the following theorem.

Theorem 1. *3-BCPP of packing 1-big 3-BCs is strongly NP-hard.*

Proof. To prove the NP-completeness, we reduce the strongly NP-complete Numerical 3-dimensional matching (3-MATCHING) [13] to the particular case of the decision version of our problem: Can all 3-BCs be packed into the Q strip cells? In the 3-MATCHING, the input data consists of three multisets X, Y, Z, each containing k positive integers, and an integer T. The sum of all elements in the sets X, Y, Z is equal to kT. Question: Does there exist a subset $M \subseteq X \times Y \times Z$, such that each number from X, Y and Z occurs exactly once, and for each triple $(x, y, z) \in M$ the equality $x + y + z = T$ holds?

Let $x_i \in X, y_i \in Y$ and $z_i \in Z$, $i = 1, \ldots, k$. Using the input data of 3-MATCHING, we construct an input data to the particular case of the 3-BCPP. Let us build k 3-BCs with the bars $a_i = 2T$, $b_i = x_i$ and $c_i = \varepsilon$, where $\varepsilon > 0$. The next k 3-BCs have bars $a_i = T + y_i$, $b_i = T - \varepsilon/2$ and $c_i = \varepsilon$. And the last k 3-BCs are defined by bars $a_i = z_i$, $b_i = T - \varepsilon/2$ and $c_i = 2T - \varepsilon$, $i = 1, \ldots, k$. Is it possible to pack such 3-BCs into the $Q = 4k$ cells? If the answer to the last question is Yes, then the answer to the question in the 3-MATCHING is Yes too. If No, then in the 3-MATCHING the answer is No too (Fig. 1). Indeed, if in the packing for some cell i the inequality $x_i + y_i + z_i < T$ holds, then for some other cell j the inequality $x_j + y_j + z_j > T$ holds since the sum of all numbers is kT. Therefore, the packing will have a length equal to $4k$ if and only if the equality $x_i + y_i + z_i = T$ holds for each $i = 1, \ldots, k$. The proof is over.

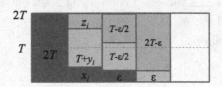

Fig. 1. Illustration to the proof of NP-hardness

4   Algorithms

The algorithms in this paper are described earlier or the modifications of the previously presented algorithms for the 2-BCPP in [8–11]. We recall and adapt these algorithms for the 3-BCPP below.

4.1   Algorithm GA

In [9] for BCPP (with arbitrary BCs), a greedy algorithm GA was proposed that can be rewritten for 3-BCPP as follows. Let a list P be an arbitrary ordered set of elements from S. The first element in P is placed in the cells $1, 2, 3$ and removed from P. Let some 3-BCs are packed and deleted from P. Items deleted from P do not move further. Then the typical procedure is performed, which consists of the following. For the next 3-BC in P, we search for the leftmost position that does not violate the packing feasibility. Its position is fixed and this 3-BC is removed from P. The algorithm stops when $P = \emptyset$. The running time of GA is $O(n^2)$ since for each 3-BC, we are searching the leftmost position in the strip.

Further in the algorithm A_3, we will apply a simplified version of the GA for packing the 1-big 3-BCs, in which the next 3-BC can participate only in 1- or 2-unions with the contents of the *last* two cells of the current packing. We will also refer to such algorithm as GA.

4.2   Algorithm A_3

Algorithm A_3 consists of three steps. At the first step, we unite some 3-BCs so that all but maybe one 3-BC become 1-big. To do this, combine a pair of 3-BCs $i, j \in S$ with bars of height a_i, b_i, c_i, a_j, b_j, $c_j \leq 1/2$ into one 3-BC with bars of the heights $a_i + a_j$, $b_i + b_j$, $c_i + c_j$. As a result, every 3-BC, except maybe one, becomes 1-big. This procedure can be performed with $O(n)$ time complexity [9].

At the second step, we split the modified set S of big 3-BCs into the six disjoint subsets as shown in Fig. 2. The set S_1 contains all big 3-BCs whose bar heights satisfy the inequalities $a \geq b \geq c$. The set S_2 includes such big 3-BCs that $a \geq c \geq b$. The set S_3 contains big 3-BCs such that $b \geq a \geq c$. The set S_4 have big 3-BCs with $c \geq b \geq a$. The set S_5 contains big 3-BCs with $c \geq a \geq b$. And the set S_6 includes big 3-BCs with $b \geq c \geq a$. If a 3-BC belongs to the several sets, then for definiteness, we place it in a set with a smaller number.

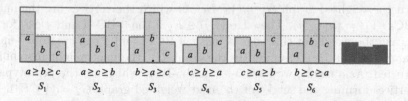

Fig. 2. Splitting big 3-BCs into the subsets

Let us arbitrarily number the elements of each set and pack the elements of each set S_1, S_2, S_3 separately using the GA from left to right, and S_4, S_5, S_6 from right to left. Since some sets S_k could be empty, we get at most six separate packings.

At the third step, we use the GA to pack the packings of sets S_1, S_2, S_3 obtained at the previous step from left to right and the packings obtained for the sets S_4, S_5, S_6 from right to left. If after the first step there is a "small" 3-BC, then we add it to the packing obtained from S_1, S_2, S_3 and apply a GA once more.

4.3   Algorithm $A_{MaxATSP(0,1)}$

In [11] an $O(n^{2.5})$-time 5/4-approximation algorithm for packing the non-increasing (or non-decreasing) big 2-BCs was proposed. The algorithm is based on the approximation-preserving reduction of 2-BCPP to the Maximum Asymmetric Traveling Salesman Problem with boolean weights of the arcs (Max-ATSP(0,1)) and using the algorithms proposed in [4,18] for the latter problem. Furthermore, in [11] an $O(n^3)$-time 16/11-approximation algorithm for packing big (not necessary non-increasing or non-decreasing) 2-BCs is presented. In obtaining this estimate, the algorithms for constructing a maximal matching [12], and one proposed in [18] are used.

If any pair of 3-BCs can build at most 1-union, we may use the same approach to find a feasible packing for 3-BCPP. Before packing, n 3-BCs occupy $3n$ cells. Each 1-union decreases the packing length by 1. Let us build a weighted complete digraph $G = (V, A)$, where the vertices are the images of the 3-BCs, and the weight of the arc $(i, j) \in A$ equals 1 if two 3-BCs i and j can form a 1-union with i on the left. Else the weight of the arc (i, j) is 0. If we find a max-weight Hamiltonian path in the graph G, the weight of this path is the number of 1-unions. The problem MaxATSP(0,1) is NP-hard. However, a 3/4-approximation $O(n^{2.5})$-time algorithm exists [4,18].

4.4   Algorithms M_w

Algorithm M_w was described in [10] for 2-BCPP. It consists of sequence of steps. For the 3-BCPP it can be described as following. First, using the set S, we construct a weighted graph $G_1 = (V_1, E_1)$, in which the vertices are the images of 3-BCs ($|V_1| = |S| = n$). The edge $(i, j) \in E_1$ if the 3-BCs i and j can form a union, and the weight of this edge equals t if BCs i and j can create a t-union, but cannot form a $(t + 1)$-union. In the graph G_1, the max-weight matching is constructed. As a result, we have 3-, 4- and 5-BCs, which are the prototypes of the vertices forming the set V_2 of the next weighted graph $G_2 = (V_2, E_2)$. The edge $(i, j) \in E_2$ of weight t exist if BCs i and j can form a t-union, but cannot form a $(t + 1)$-union. At an arbitrary step in the corresponding graph G_k, we construct the next max-weight matching. The algorithm stops when in the next graph G_{k+1}, there are no more edges.

5   Approximation Results

The following theorem is valid for packing *arbitrary* 3-BCs.

Theorem 2. *Algorithm A_3 with time complexity $O(n)$ constructs a packing for 3-BCPP of length at most $3 \cdot OPT + 2$, where OPT is the minimum length of the strip into which n 3-BCs can be packed.*

Proof. Since the sets S_1, S_2, S_3 and S_4, S_5, S_6 are symmetric, we consider in detail only packing of the sets S_1, S_2, S_3.

All 3-BCs of the sets $S_1, \ldots, S_6$ are 1-big by construction, which means that each has a bar higher than $1/2$. One 3-BC is packed in the first three cells. Therefore, the packing density of one of the occupied cells is more than $1/2 > 1/3$.

Suppose that after packing k 3-BCs, the packing density of all cells, except two, is at least $1/3$.

Let us consider what happens when the $(k+1)$-th 3-BC in the S_1 is packed using a simplified version of the GA. At this point, k 3-BCs are already packed and form a BC, which we denote as $\overline{S}_1^k$. Evidently, $a_{k+1} + a_k > 1$. Let y_k and z_k are the heights of the two last bars in the $\overline{S}_1^k$. The following three cases are possible.

1. $a_{k+1} + y_k < 1$ (the first bar of the $(k+1)$-th 3-BC shares a cell with the last but one bar of the $\overline{S}_1^k$). Then, the packing density of all cells, except the last one, is greater than $1/2 > 1/3$ (because $a_{k+1} > 1/2$).
2. $a_{k+1} + y_k > 1$, but $a_{k+1} + z_k < 1$ (the first bar of the $(k+1)$-th 3-BC cannot be united with the last but one bar of the $\overline{S}_1^k$, but the first bar of the $(k+1)$-th 3-BC can be united with the last bar of the $\overline{S}_1^k$). Then, $y_k + a_{k+1} + z_k > 1$ (because $y_k + a_{k+1} > 1$). Therefore, the density of the last but one and third cell from the end is greater than $1/2 > 1/3$.
3. $a_{k+1} + y_k > 1$ and $a_{k+1} + z_k > 1$ ($\overline{S}_1^k$ and $(k+1)$-th 3-BCs cannot be united). In this case $a_{k+1} + y_k + z_k > 1$, and it follows that the packing density of all cells, except the last two, is greater than $1/3$.

Therefore, after packing $(k+1)$-th 3-BC the packing density of all cells, except two, is greater than $1/3$.

Let us consider the packing of the set S_2. We suppose that after packing k 3-BCs, the packing density of all cells, except two, is greater than $1/3$. Let $\overline{S}_2^k$ be the BC, which is the result of packing k 3-BCs, y_k is the height of the last but one and z_k is a height of its last bar. The following cases are possible when packing the $(k+1)$-th 3-BC.

1. If $a_{k+1} + y_k < 1$, $a_{k+1} + z_k < 1$, then we pack the first and the second bars of the $(k+1)$-th 3-BC in the cells containing two last bars of the $\overline{S}_2^k$ (the sum of heights of these bars is less than 1 since $a > b$). Since $a_{k+1} > 1/2 > 1/3$, we get the packing density greater than $1/3$ except the last two cells.
2. If $a_{k+1} + y_k < 1$, $a_{k+1} + z_k > 1$, but $b_{k+1} + z_k < 1$, then we pack the first and the second bars of the $(k+1)$-th 3-BC in the cells containing two last bars of the $\overline{S}_2^k$. We get the packing density greater than $1/3$ except the last two cells, since $a_{k+1} > 1/2 > 1/3$.

3. If $a_{k+1}+y_k > 1$, but $a_{k+1}+z_k < 1$, then we unite the first bar of the $(k+1)$-th 3-BC and the last bar of the $\overline{S}_2^k$. Since $a_{k+1} + y_k > 1$, the packing density is more than $1/3$, except maybe the last 2 cells.

4. If $(a_{k+1} + y_k > 1$ and $a_{k+1} + z_k > 1)$ or $(a_{k+1} + y_k < 1$ and $b_{k+1} + z_k < 1)$, then we pack the $(k + 1)$-th 3-BC on the right into the empty cells. We get that $a_{k+1} + z_k > b_{k+1} + z_k > 1$, whence it follows that the packing density, not counting the last two cells, is greater than $1/3$.

Let us consider the packing of the set S_3. We suppose that k 3-BCs are packed with density greater than $1/3$ except two cells. Let us define the result of packing k 3-BCs as BC $\overline{S}_3^k$, where y_k is the height of the last but one and z_k is the height of the last bar. The following cases arise when packing the $(k+1)$-th 3-BC.

1. If $a_{k+1}+y_k < 1$ and $b_{k+1}+z_k < 1$, then we unite the first bar of the $(k+1)$-th 3-BC and the last but one bar of the $\overline{S}_3^k$. We have $b_{k+1} + z_k > 1/2 > 1/3$. Therefore, the packing density of all cells except one is greater than $1/3$.

2. If $a_{k+1}+y_k < 1$ and $b_{k+1}+z_k > 1$, then we unite the first bar of the $(k+1)$-th 3-BC and the last bar of the $\overline{S}_3^k$. We have $b_{k+1} + z_k > 1$, which means that the packing density of all cells is greater than $1/3$.

3. If $a_{k+1} + y_k > 1$, $a_{k+1} + z_k > 1$, then we pack the $(k + 1)$-th 3-BC into the empty cells. Then $a_{k+1} + z_k + b_{k+1} > 1 + 1/2 = 3/2$, that is, the packing density of all cells is greater than $1/3$.

4. If $a_{k+1} + y_k > 1$, $a_{k+1} + z_k < 1$, $b_{k+1} + z_k > 1$, then we unite the first bar of the $(k+1)$-th 3-BC and the last bar of the $\overline{S}_3^k$. We have $b_{k+1} + z_k > 1$, which means that the packing density of all cells is greater than $1/3$.

In the latter case $a_{k+1}+b_k+z_{k-1} = a_{k+1}+y_k > 1$, $a_{k+1}+c_k < 1$, $b_{k+1}+c_k < 1$, so we unit the first bar of the $(k+1)$-th 3-BC and the last bar of the $\overline{S}_3^k$, but we cannot use the height of the y_k bar, because it could affect the packing density of early filled cells. To show that the density remains greater than $1/3$, we consider the options for packing the k-th 3-BC together with the $(k + 1)$-th 3-BC. Note that in the cases 2–4, the packing density is greater than $1/3$, including the cell containing the bar of height z_{k+1}. Therefore, we can divide the packing cases of the k-th 3-BC into 3 variants (including new cases that will appear due to this technique):

   – k-th 3-BC is packed in the same way as in the case 1 or it is the first 3-BC in the S_3;
   – k-th 3-BC is packed in the same way as in the cases 2–5, 7;
   – k-th 3-BC is packed in the same way as in the case 6.

5. If $a_{k+1} + b_k + z_{k-1} = a_{k+1} + y_k > 1$, $a_{k+1} + c_k < 1$, $b_{k+1} + c_k < 1$ and the k-th 3-BC is packed in the same way as in the case 1, then we unite the first bar of the $(k + 1)$-th 3-BC and the third bar of the k-th 3-BC. Since $a_{k+1} + b_k + b_{k+1} > 1 + 1/2 = 3/2$, then the density of the four last cells is greater than $1/3$. Note that in this case, the cell containing the bar of height c_{k+1} has a density greater than $1/3$, as in the cases 2–4.

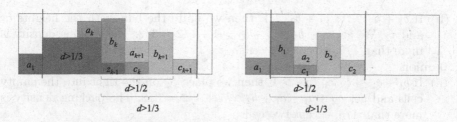

Fig. 3. A case 5 for the set S_3 packing

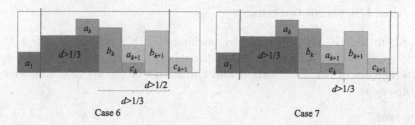

Fig. 4. The cases 6 and 7 for the set S_3 packing

6. If $a_{k+1} + b_k > 1$, $a_{k+1} + c_k < 1$, $b_{k+1} + c_k < 1$ and the k-th 3-BC is packed in the same way as in the case 2, then we unite the bars with heights a_{k+1} and c_k. We get that $b_{k+1} > 1/2 > 1/3$, and the packing density is greater than $1/3$, except one cell.

7. If $a_{k+1} + b_k > 1$, $a_{k+1} + c_k < 1$, $b_{k+1} + c_k < 1$, then we unite a_{k+1} and c_k. Since $a_{k+1} + b_k + c_k + b_{k+1} > 1 + 1/2 = 3/2$, the density of the four last cells is greater than $1/3$. Note that in this case, the cell containing the bar of height c_{k+1} has density greater than $1/3$, as in the cases 2–5.

When packing the 3-BCs of the set S_3, the height of the first bar a_1 of the first 3-BC can be arbitrary. Thus, the packing density of all cells, except two, is more than $1/3$.

After applying the GA to all types of the 3-BCs (the sets S_4, S_5 and S_6 are packed using a simplified version of the GA from right to left), we get six packings, which we denote as $\overline{S}_1, \ldots, \overline{S}_6$. These packings can be considered as new BCs and can also be packed. Let us consider the BCs $\overline{S}_1$ and $\overline{S}_2$. Denote by b_1 and c_1 the heights of the last but one and the last bars of $\overline{S}_1$, and let a_2 and b_2 are the heights of the first and the second bars of $\overline{S}_2$. When packing $\overline{S}_1$ and $\overline{S}_2$, we need to consider the following cases.

1. 2-union.
   If $b_1 + a_2 < 1$, $c_1 + b_2 < 1$, then we unite the corresponding bars and get a packing with density greater than $1/3$, except two cells.

2. 1-union.
   (a) If $b_1 + a_2 < 1$, $c_1 + b_2 > 1$, then we unite the bars with the heights c_1 and a_2. We get $a_2 + b_1 + c_1 + b_2 > 1 + 1/2 = 3/2$. The packing density is more than $1/3$, except two cells.

(b) If $b_1 + a_2 > 1$, $c_1 + b_2 < 1$, then we unite the bars with the heights c_1 and a_2. We get $a_2 + b_1 + c_1 + b_2 > 1 + 1/2 = 3/2$. The packing density is more than $1/3$, except two cells.

3. 0-union.
   (a) If $b_1 + a_2 > 1$, $c_1 + b_2 > 1$, then we place S_2 on the right into the empty cells and get $a_2 + b_1 + c_1 + b_2 > 1 + 1/2 = 3/2$. The packing density is more than $1/3$, except two cells.
   (b) If $c_1 + b_2 > 1$, $c_1 + a_2 > 1$, then we put S_2 on the right without union and get $a_2 + b_1 + c_1 + b_2 > 1 + 1/2 = 3/2$. The packing density is more than $1/3$, except two cells.

Next, to the BC $\overline{S}_{1,2}$ obtained by packing the BCs $\overline{S}_1$ and $\overline{S}_2$, we add $\overline{S}_3$. Let a_1, b_1, c_1 are the heights of the last three bars of the BC $\overline{S}_{1,2}$, and a_2, b_2, c_2 are the heights of the first, second and third bars of the BC $\overline{S}_3$. Consider the possible cases:

1. 3-union.
   $a_1 + a_2 < 1$, $b_1 + b_2 < 1$, $c_1 + c_2 < 1$, then we unite the corresponding bars and get a packing with density greater than $1/3$, except two cells.
2. 2-union.
   (a) If $a_1 + a_2 > 1$, $b_1 + a_2 < 1$, $c_1 + b_2 < 1$.
       Then we unite the corresponding bars and get a packing having a density greater than $1/3$, with the exception of two cells. One of these cells (with density less than $1/3$) is the last one in the current packing.
   (b) If $b_1 + b_2 > 1$, $b_1 + a_2 < 1$, $c_1 + b_2 < 1$.
       In this case we unite the corresponding bars and get a packing with density greater than $1/3$, except two cells. One of these cells is the last one in the current packing.
   (c) If $c_1 + c_2 > 1$, $b_1 + a_2 < 1$, $c_1 + b_2 < 1$.
       We unite the corresponding bars and get a packing with density greater than $1/3$, except for two cells. One of these cells is the last one in the current packing.
3. 1-union.
   (a) If $a_1 + a_2 > 1$, $b_1 + a_2 > 1$, $c_1 + a_2 < 1$.
       We unite the corresponding bars. Since $b_1 + a_2 > 1$, we get a packing with density greater than $1/3$, except one cell.
   (b) If $a_1 + a_2 > 1$, $c_1 + b_2 > 1$, $c_1 + a_2 < 1$.
       Bar $b2$ of the first 3-BC in $\overline{S}_3$ for all cases of packing of the set S_3 is responsible only for the cell in which it is located, except case 5. In this case, its height affects the density of subsequent cells, and we cannot use the inequality $c_1 + b_2 > 1$ to get the needed density. It is required to consider how the last 3-BC from the BC $\overline{S}_{1,2}$ unite the first 3-BC in $\overline{S}_3$. These may be 3 cases for the set S_1 or S_2 considered earlier, or the BC $\overline{S}_{1,2}$ contains only one 3-BC. There are seven cases in total:
       Let a_1, b_1, c_1 are the heights of the bars of the last but one 3-BC from $\overline{S}_{1,2}$, a_2, b_2, c_2 are the heights of the bars of the last 3-BC from $\overline{S}_{1,2}$,

a_3, b_3, c_3 are the heights of the bars of the first 3-BC from S_3, a_4, b_4, c_4 are the heights of the bars of the second 3-BC from $\overline{S}_3$.

Then $a_2 + a_3 > 1$, $c_2 + b_3 > 1$, $c_2 + a_3 < 1$.

   i. Case 1 for packing $\overline{S}_1$.

      We have that $a_2 + a_3 + c_2 + b_3 + b_4 > 1 + 1 + 1/2 = 2.5$. Therefore, the density of all cells, except one, is greater than $1/3$.

   ii. Case 2 for packing $\overline{S}_1$.

      In this case $b_1 + a_2 + c_2 + b_3 + b_4 > 1 + 1 + 1/2 = 2.5$. Then the density of all cells except one is greater than $1/3$.

   iii. Case 3 for packing $\overline{S}_1$.

      In this case $b_1 + a_2 + c_2 + b_3 + b_4 > 1 + 1 + 1/2 = 2.5$, and the density of all cells, except two, is greater than $1/3$. Where one of these cells (with density less than $1/3$) is the last one in the current packing.

   iv. Case 1 for packing $\overline{S}_2$.

      Similarly to the case 1 for $\overline{S}_1$.

   v. Case 2 for packing $\overline{S}_2$.

      Similarly to the case 2 for $\overline{S}_1$.

   vi. Case 2 for packing $\overline{S}_2$.

      Similarly to the case 3 for $\overline{S}_1$.

   vii. Let S_1 or S_2 consists of one 3-BC.

      Since $\overline{S}_1$ or $\overline{S}_2$ is the first 3-BC, then $a_2 + a_3 + b_3 + a_4 + b_4 > 1 + 1 + 1/2 = 2.5$. Therefore, the density of all cells except one is greater than $1/3$.

(c) If $b_1 + b_2 > 1$, $c_1 + b_2 > 1$, $c_1 + a_2 < 1$.

    Similar to the previous case, but when there is one 3-BC in the $\overline{S}_1$ or $\overline{S}_2$, we have that $a_2 + b_3 + a_4 + b_4 > 1/2 + 1 + 1/2 = 2$, and the density of all cells, except two, is greater than $1/3$. And one of the cells with density less than $1/3$ is inside the BC.

(d) If $c_1 + c_2 > 1$, $c_1 + b_2 > 1$, $c_1 + a_2 < 1$.

    In this variant, in case 5 of uniting the first 3-BC in $\overline{S}_3$, we can use the height of the bar c_3, so we have $c_1 + c_2 > 1$, and the density of all cells, except one, is greater than $1/3$.

4. 0-union.

   If $a_2 + a_1 > 1$, $a_2 + b_1 > 1$, $a_2 + c_1 > 1$, then since $a_2 + c_1 > 1$ the density of all cells except one is greater than $1/3$.

Similarly, we construct the BC packing $\overline{S}_4, \overline{S}_5, \overline{S}_6$ from right to left, and also get a packing density more than $1/3$, except two cells.

A complete proof with illustrations one can find in [1].

We used a simplified version of the GA, where for each BC we consider only the unions with the last 2 bars. Then the time complexity of algorithm A_3 is $O(n)$. The theorem is proved.

For the particular case when all 3-BCs are 2-big, we prove the following new results.

Lemma 1. *If all 3-BCs are 2-big, then $OPT \geq 2n$.*

Proof. More than one big bar cannot be in one cell, so each 2-big 3-BC occupies at least 2 cells. Since $|S| = n$, the optimal packing length is at least $2n$. The lemma is proved.

Theorem 3. *If all 3-BCs are 2-big, and the second bar of each 3-BC is big, then algorithm $A_{MaxATSP(0,1)}$ constructs a 9/8-approximate solution with time complexity $O(n^{2.5})$.*

Proof. Since the first two or last two bars of each 3-BC are higher than $1/2$, only 1-unions are possible. Moreover, $OPT \geq 2n$ (Lemma 1). If we solve the $MaxATSP(0,1)$ using approximation algorithm [18] with approximation ratio $3/4$ (i.e. $q \geq \frac{3}{4}q^*$, where q^* is the optimum of the $MaxATSP(0,1)$, and q is the weight of the path constructed by approximation algorithm), then we get an approximate packing of length at most $3n - q$. From the inequality $q^* \leq n - 1$ follows

$$\frac{3n - q}{3n - q^*} \leq \frac{3n - \frac{3}{4}q^*}{3n - q^*} \leq \frac{9}{8}.$$

The theorem is proved.

Let now each 3-BC is arbitrary 2-big (any two bars of each BC are higher than $1/2$). Algorithm M_w constructs a max-weight matching at each step. We can assume that the optimal algorithm also builds some matching at each step [10]. Let us introduce the following notation.

- m_k is the cardinality of the k-th matching constructed by the M_w;
- m_k^* is the cardinality of k-th matching in the optimal packing;
- w_k is the weight of the matching constructed at the k-th step of the M_w;
- w_k^* is the weight of the matching constructed at step k in the optimal packing;
- k_1 is the number of 2-unions in the first matching constructed by the M_w;
- k_1^* is the number of 2-unions in the first matching in the optimal packing.

Lemma 2. *Any packing of 2-big 3-BCs can be disassembled so that 2-unions appear only in the first matching.*

Proof. Suppose we have a packing P. Let us disassemble it into the separate 3-BCs that are not involved in the 2-unions (set A), and into the 4-BCs obtained as a result of 2-unions (set B). We need to show that 2-unions can appear only in the first matching.

We disassemble the packing P from left to right. Consider the first 3-BC. If it is united with another BCs by:

1. 0-union or 1-union. Then we exclude it from the packing P and put it in the set A;
2. 2-union. Then we consider such a 4-BC, which is the result of a 2-union. This 4-BC can only has 0- and 1- unions with other BCs. Let us exclude it from P and include in the set B.

Repeating this procedure until P becomes empty, we get a set A of separate 3-BCs, and a set B of 4-BCs obtained after 2-unions. Since each 2-big 3-BC can participate in a 2-union at most once, we assume that the first matching constructs all 2-unions. The lemma is proved.

The Lemma implies the equalities $w_k^* = m_k^*, \ k \geq 2$.

Lemma 3. *If all 3-BCs are 2-big, then $m_2^* + m_3^* + \ldots + m_q^* \leq m_1^*$.*

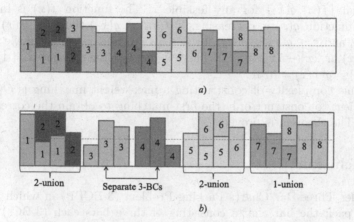

Fig. 5. Disassembling of the packing. a) Optimal packing; b) First matching

Proof. Since the cardinality of the first matching in the optimal packing is m_1^*, then $B = n - 2m_1^*$ 3-BCs cannot form either 2- or 1-unions with each other. Let us call them the "separate" 3-BCs (in the Fig. 5 $m_1^* = 3$, $B = 2$). In the following matchings, starting from the second, these B separate 3-BCs can only form 1-unions with the m_1^* 4-BCs resulting from the unions in the first step. Furthermore, m_1^* new 4-BCs can also be united with each other. However, in total, starting from the second step, at most $m_1^* - 1$ 1-unions are possible since B 3-BCs do not unite with each other. The lemma is proved.

Theorem 4. *If all 3-BCs are 2-big, then algorithm M_w constructs a 5/4-approximate solution with time complexity $O(n^3)$.*

Proof. Initially, n 3-BCs occupy $3n$ cells of the strip. Each 1-union reduces the packing length by 1, and 2-union reduces by 2. If p matchings are constructed, then the length of the packing yielded by the M_w is

$$L(M_w) = 3n - w_1 - w_2 - \ldots - w_p \leq 3n - w_1.$$

From Lemma 2 and Lemma 3 follows

$$OPT = 3n - w_1^* - w_2^* - \ldots - w_q^* = 3n - w_1^* - m_2^* - \ldots - m_q^* \geq 3n - w_1^* - m_1^*.$$

Then

$$\varepsilon = \frac{L(M_w)}{OPT} \leq \frac{3n - w_1}{3n - w_1^* - m_1^*} \leq \frac{3n - w_1}{3n - w_1 - n/2} = 1 + \frac{1}{5 - 2w_1/n} = f(x),$$

where $x = w_1/n$. On the other hand, Lemma 1 implies

$$\varepsilon = \frac{L(M_w)}{OPT} \leq \frac{3n - w_1}{2n} = \frac{3 - w_1/n}{2} = g(x).$$

So, $\varepsilon \leq \min\{f(x), g(x)\}$ for any feasible x. The function $f(x)$ is increasing, while the function $g(x)$ is decreasing. If $f(x_0) = g(x_0)$ then $\varepsilon \leq f(x) \leq f(x_0)$ for $x \leq x_0$ and $\varepsilon \leq g(x) \leq g(x_0)$ for $x \geq x_0$. To find x_0 we solve the equation $f(x) = g(x)$ or $2x^2 - 7x + 3 = 0$. There is one feasible solution $x = 1/2$. Then $\varepsilon \leq f(1/2) = g(1/2) = 5/4$.

The time complexity of constructing a max-weight matching is $O(n^3)$ [12]. It is sufficient to construct only the first matching to obtain the corresponding accuracy. The theorem is proved.

6 Conclusion

We consider Three-Bar Charts Packing Problem (3-BCPP) in which it is necessary to pack the bar charts consisting of three bars each (3-BCs) into the horizontal unit-height strip of minimal length. The bars of each 3-BC may move vertically within the strip, but it is forbidden to change the order and separate the bars of each 3-BC. This novel issue is a generalization of the strongly NP-hard Two-Bar Charts Packing Problem (2-BCPP) considered earlier [9–11].

- We proposed an $O(n)$-time algorithm A_3 which constructs a packing of length at most $3 \cdot OPT + 2$, where OPT is the optimum of 3-BCPP.
- If all 3-BCs are 2-big (at least two bars have a height more than $1/2$), we show how to find a 5/4-approximate solution with time complexity $O(n^3)$. If, additionally, a small bar of each 3-BCs is not a second, it is possible to find a 9/8-approximate packing with time complexity $O(n^{2.5})$.
- If all 3-BCs are 1-big (one bar of each 3-BC is higher than $1/2$), we prove that 3-BCPP remains strongly NP-hard. The complexity of packing 2-big 3-BCs we do not know yet.

In the future, we are planning to prove that A_3 constructs a packing of length at most $8/3\ OPT + 1$. Moreover, we want to show the tightness of the obtained accuracy estimates of the considered algorithms.

References

1. Complete Proof of Theorem 2: https://drive.google.com/drive/folders/1SLUye20gBVaOWI4YM8z72sgDytJTHs63?usp=sharing

2. Baker, B.S.: A new proof for the first-fit decreasing bin-packing algorithm. J. Algorithms **6**, 49–70 (1985)
3. Bansal, N., Eliás, M., Khan, A.: Improved approximation for vector bin packing. In: SODA, pp. 1561–1579 (2016)
4. Bläser, M.: A 3/4-approximation algorithm for maximum ATSP with weights zero and one. In: Jansen, K., Khanna, S., Rolim, J.D.P., Ron, D. (eds.) APPROX/RANDOM -2004. LNCS, vol. 3122, pp. 61–71. Springer, Heidelberg (2004). https://doi.org/10.1007/978-3-540-27821-4_6
5. Blaser, M.: An 8/13-approximation algorithm for the maximum asymmetric TSP. J. Algorithms **50**(1), 23–48 (2004)
6. Christensen, H.I., Khanb, A., Pokutta, S., Tetali, P.: Approximation and online algorithms for multidimensional bin packing: a survey. Comput. Sci. Rev. **24**, 63–79 (2017)
7. Dósa, G.: The tight bound of first fit decreasing bin-packing algorithm is $FFD(I) \leq 11/9\,OPT(I) + 6/9$. In: Chen, B., Paterson, M., Zhang, G. (eds.) ESCAPE 2007. LNCS, vol. 4614, pp. 1–11. Springer, Heidelberg (2007). https://doi.org/10.1007/978-3-540-74450-4_1
8. Erzin, A., Plotnikov, R., Korobkin, A., Melidi, G., Nazarenko, S.: Optimal investment in the development of oil and gas field. In: Kochetov, Y., Bykadorov, I., Gruzdeva, T. (eds.) MOTOR 2020. CCIS, vol. 1275, pp. 336–349. Springer, Cham (2020). https://doi.org/10.1007/978-3-030-58657-7_27
9. Erzin, A., Melidi, G., Nazarenko, S., Plotnikov, R.: Two-bar charts packing problem. Optim. Lett. **15**(6), 1955–1971 (2021)
10. Erzin, A., Melidi, G., Nazarenko, S., Plotnikov, R.: A 3/2-approximation for big two-bar charts packing. J. Comb. Optim. **42**(1), 71–84 (2021). https://doi.org/10.1007/s10878-021-00741-1
11. Erzin, A., Shenmaier, V.: An improved approximation for packing big two-bar charts. J. Math. Sci. **267**(4), 465–473 (2022). https://doi.org/10.1007/s10958-022-06151-w
12. Gabow, H.: An efficient reduction technique for degree-constrained subgraph and bidirected network flow problems. In: STOC, pp. 448–456 (1983)
13. Garey, M.R., David, S.J.: Computers and Intractability; A Guide to the Theory of NP-Completeness. p. 339 (1979)
14. Johnson, D.S.: Near-optimal bin packing algorithms. Ph.D thesis, Massachusetts Institute of Technology (1973)
15. Johnson, D.S., Garey, M.R.: A 71/60 theorem for bin packing. J. Complex. **1**(1), 65–106 (1985)
16. Kellerer, H., Kotov, V.: An approximation algorithm with absolute worst-case performance ratio 2 for two-dimensional vector packing. Oper. Res. Lett. **31**, 35–41 (2003)
17. Li, R., Yue, M.: The proof of $FFD(L) \leq 11/9\,OPT(L) + 7/9$. Chin. Sci. Bull. **42**(15), 1262–1265 (1997)
18. Paluch, K.: Maximum ATSP with weights zero and one via half-edges. Theory Comput. Syst. **62**(2), 319–336 (2018)
19. Vazirani, V.V.: Approximation Algorithms. Springer, Heidelberg (2003). https://doi.org/10.1007/978-3-662-04565-7
20. Yue, M.: A simple proof of the inequality $FFD(L) \leq 11/9\,OPT(L) + 1, \forall L$, for the FFD bin-packing algorithm. Acta Math. Appl. Sin. **7**(4), 321–331 (1991)
21. Yue, M., Zhang, L.: A simple proof of the inequality $MFFD(L) \leq 71/60\,OPT(L) + 1, \forall L$, for the MFFD bin-packing algorithm. Acta Math. Appl. Sin. **11**(3), 318–330 (1995)

An 11/7 — Approximation Algorithm for Single Machine Scheduling Problem with Release and Delivery Times

Natalia Grigoreva$^{(\boxtimes)}$ (iD)

St. Petersburg State University, Universitetskaja nab. 7/9,
199034 St. Petersburg, Russia
n.s.grig@gmail.com

Abstract. In this paper we consider the single machine scheduling problem. Each job has a release time, processing time and a delivery time. Preemption of jobs is not allowed. The objective is to minimize the time, by which all jobs are delivered. This problem is denoted by $1|r_j, q_j|C_{\max}$, has many applications, and it is NP-hard in strong sense. The problem is useful in solving flowshop and jobshop scheduling problems. The goal of this paper is to propose a new 11/7— approximation algorithm, which runs in $O(n \log n)$ times. To compare the effectiveness of proposed algorithms we tested random generated problems.

Keywords: Single machine scheduling problem · Inserted idle time · Worst-case performance ratio · Approximation algorithm

1 Introduction

The single machine problem of minimizing the maximum delivery times $1|r_j, q_j|C_{\max}$, (Graham *et al.* [1]) is a classical combinatorial optimization problem. We consider a set of jobs $V = \{1, 2, \ldots, n\}$. Each job i must be processed without interruption for $t(i)$ time units on the processor, which can process at most one job at time. Each job i has a release time $r(i)$, when the job is ready for processing, and a delivery time $q(i)$. With no loss of generality we consider that $t(i), r(i), q(i)$ are integers. The delivery of each job begins immediately after processing has been completed. The objective is to minimize the time, by which all jobs are delivered.

It is required to construct a schedule, that is, to find for each job $i \in V$ the start time $\tau(i)$, provided that $r(i) \leq \tau(i)$. We can construct the permutation of jobs $S = (i_1, i_2, \ldots, i_n)$ and then find the start time $\tau(i)$ for each job by formula $\tau(i_j) = \max\{r(i_j), \tau(i_{j-1}) + t(i_{j-1})\}$, for $j \geq 2$ and $\tau(i_1) = r(i_1)$.

The goal is to construct a schedule that minimizes the delivery time of the last job $C_{\max} = \max\{\tau(i) + t(i) + q(i)|i \in V\}$. The problem is NP-hard in the strong sense [2], but there are exact polynomial algorithms for some special cases and has many applications.

N. Olenev et al. (Eds.): OPTIMA 2022, CCIS 1739, pp. 76–89, 2022.
https://doi.org/10.1007/978-3-031-22990-9_6

Some authors considered an equivalent formulation of the problem, in which instead of the delivery time for each job, the due date $D(i) = K - q(i)$ is known, where K is a constant, and the objective function is the maximum lateness $L_{\max} = \max\{\tau(i) + t(i) - D(i) | i \in V\}$. This formulation of the problem is denoted as $1|r_i|L_{\max}$.

If we swap the delivery times and the release times, we get an inverse problem with the property that the solution of the direct problem $S = (i_1, i_2, ..., i_n)$ is optimal if and only if the permutation $S_{inv} = (i_n, i_{n-1}, ..., i_1)$ is the optimal solution of the inverse problem.

The problem $1|r_j, q_j|C_{\max}$ is the main subproblem in many important models of scheduling theory, such as multiprocessor scheduling, flowshop and jobshop problems. The study of this problem has the theoretical interest and it is useful in practical industrial applications [3–5,14].

Among the early works [10–13], that developed branch and bound algorithms for single processor scheduling problem, the most effective algorithm was Carlier's algorithm. This algorithm constructs a full solution by extended Jackson's rule in each node of the search tree and optimally solves random instances with 50–10,000 jobs. But there are hard instances for Carlier's algorithm.

In further works, authors improve Carlier's algorithm by offering various methods for obtaining lower and upper bounds.

One way to improve the performance of the branch and bound method is to use approximation efficient algorithms to obtain upper bounds. Such algorithms should have a good approximation ratio and the low computational complexity.

In [14] three branch and bound algorithms based on Carlier's algorithm were presented. The paper [15] considered the the single processor scheduling problem with precedence constraints and proposed the branch and bound algorithm which used three different heuristics at each branch node.

The branch and bound algorithms proposed in [4], used a binary branching rule where at each branch node the full schedule is generated by new heuristic algorithm.

We developed the branch and bound algorithm for the single processor scheduling problem with precedence constraints [18]. The branch and bound algorithm used two heuristics at each branch node: extended Jackson's rule and an inserted idle time algorithm IJR.

One of the popular scheduling tools are list algorithms that build non-delayed schedules. In the list algorithm, at each step, the job with the highest priority is selected from the set of ready jobs. But the optimal schedule may not belong to the class of non-delayed schedules.

IIT (inserted idle time) schedules were defined in [16] as feasible schedules in which the processor can be idle when there are jobs ready to run.

The main idea of greedy algorithms for solving these problems is the choice at each step of the highest priority job, before the execution of which the processor could be idle.

Several approximation algorithms are known for solving the problem $1|r_i, q_i|C_{\max}$, for which the worst-case performance ratio is established. The

Table 1. Approximation algorithms

Year	P_n	Comp. c	WCR	Ref.
1971	1	$O(n \log n)$	2	[6]
1980	n	$O(n^2 \log n)$	3/2	[7]
1992	4n	$O(n^2 \log n)$	4/3	[8]
1994	2	$O(n \log n)$	3/2	[9]
2021	2	$O(n \log n)$	3/2	[17]

Table 1 contents information about its. In the first column there is the year of publication, second column P_n shows a number of permutations, which the algorithm constructs, computational complexity of algorithm is given in the 3 column, the worst-case performance ratio is given in the 4 column, in the last column references are given.

The first algorithm for constructing an approximate schedule was the Schrage heuristic [6] - an extended Jackson rule, which is formulated as follows: each time the processor is free, a ready job with the maximum delivery time is assigned to it.

K. Potts [7] proposed an algorithm in which the extended Jackson's rule algorithm repeats n times. The best schedule is selected from n constructed schedules.

L. Hall and D. Schmois [8] considered the direct problem and the inverse problem and have developed a method in which the Potts algorithm is applied to the direct and the inverse problem. In total, the algorithm builds $4n$ schedules and chooses the best one.

E. Novitsky and K. Smutnitsky [9] proposed an $3/2$ — approximation algorithm, which creates only two permutations. For the first time, the Jackson rule is applied, then the interference work is determined and the set of tasks is divided into two sets: tasks that should be performed before the interference work and tasks that should be performed after it. The best schedule is selected from two schedule.

The author proposed an IIT $3/2$—approximation algorithm ICA for a single-machine scheduling problem [17], which creates two permutations. One permutation is constructed by the Jackson rule algorithm, other permutation is constructed by IJR (Idle Jackson Rule) algorithm. We select the best one.

The goal of the paper is to prepare an approximation IJR algorithm, that builds only one permutation and to prove that the worst-case performance ratio of the IJR algorithm is equal $11/7$.

Schrage heuristic builds one permutation and has a guaranteed accuracy score of 2. All other algorithms actively use this algorithm. The proposed algorithm has the best guaranteed estimate and can be used to construct other algorithms with the best guaranteed estimate. For example, it is interesting to apply this algorithm to direct and inverse problems (as in the method [8]). In addition, the algorithm is twice as fast as algorithms [9] and [17] because it only builds one permutation.

The article is organized as follows: Sect. 2 presents an approximation algorithm IJR. The theoretical study of the proposed algorithm in Sect. 3 contains three lemmas and one theorem. We prove that the worst-case performance ratio of the IJR algorithm is equal 11/7. The results of the computational experiment are given in Sect. 4. In conclusion, the main results obtained in the article are formulated.

2 IJR Scheduling Algorithm

The main idea of the IJR algorithm is that sometimes it is better to place a priority job on service, even if it leads to some idle time of the processor.

In the IJR algorithm two jobs are selected: the highest priority job and the highest priority ready job. The paper has established special conditions in which it is advantageous to organize the unforced idle time of the processor. These conditions allow to choose between two jobs.

The algorithm IJR is a greedy algorithm, but not a list algorithm and can be used as a basic heuristic for various scheduling models and constructing a branch and bound method.

We introduce the following notation: $S_k = (i_1, i_2, \ldots, i_k)$ is the partial schedule, $time := \max\{\tau(i) + t(i) | i \in S_{k-1}\}$ is the time to release the processor after the execution of already scheduled jobs. We store ready jobs in the queue with priorities Q_1, the priority of a job is its delivery time.

2.1 Algorithm IJR

Initialization

1. Sort all tasks in non-descending order of receive times:
   $r(j_1) \leq r(j_2) \leq \ldots \leq r(j_n)$.
   Let the list $H = (j_1, j_2, \ldots, j_n)$
2. Define $r_{min} = \min\{r(i) | i \in V\}$.
3. Define $q_{min} = \min\{q(i) | i \in V\}$.
4. Define two lower bounds the objective function
   $LB_1 = r_{min} + \sum_{i=1}^{n} t(i) + q_{min}$.
   $LB_2 = \max\{r(i) + t(i) + q(i) | i \in V\}$.
5. Define the lower bound of the objective function $LB = \max\{LB_1, LB_2\}$.
6. Set $time := r_{min}$. $Q_1 = \emptyset$, $l = 1$, $S_0 = \emptyset$.

   Main loop
   For $k = 1$ to n do

1. while $(r(j_l) \leq time)$ do
   begin add a ready job $r(j_l)$ to the Q_1 queue; l:=l+1 end.
2. If there is no ready job and $Q_1 = \emptyset$, then
   $time := \min\{r(i) | i \notin S_{k-1}\}$ and go to step 1.

3. Select the ready job $u \in Q_1$ with the maximum delivery time
   $q(u) = \max\{q(i)|i \in Q_1\}$.
4. Set $r_{up} := time + t(u)$.
5. While there are jobs j_l such that $time < r(j_l) < r_{up}$ do
   (a) If $q(j_l) \leq LB/2$, then j_l is added to the queue Q_1. $l := l+1$, go to step 5.
   (b) If $q(j_l) > LB/2$, define a possible idle time of the processor $idle(j_l) = r(j_l) - time$.
   (c) If $q(j_l) - q(u) \geq idle(j_l)$ then we set task j_l on the processor:
       set $\tau(j_l) := r(j_l)$; $time := \tau(j_l) + t(j_l)$,
       $S_k := S_{k-1} \cup \{j_l\}$, $l := l+1$, go to step 1.
   (d) else Add task j_l, to the queue Q_1.
       Set $l := l+1$, go to step 5.
6. Set the job u on the processor:
   set $\tau(u) := time$; $time := \tau(u) + t(u)$,
   $S_k := S_{k-1} \cup \{u\}$.
   Delete u from the queue Q_1, go to step 1.
7. The schedule S_n is constructed. Find the value of the objective function
   $C_{\max}(S_n) = \max\{\tau(i) + t(i) + q(i) \mid i \in V\}$.

The algorithm sets on the processor the job j_l with the large delivery time $q(j_l)$. If this job is not ready, then the processor will be idle in the interval $[time, r(j_l)]$. To avoid too much idle of the processor the inequality $q(jl) - q(u) \geq idle(j_l)$ is verified on step 5c. If the inequality is hold, we choose job j_l. Otherwise, we set the ready job u with the maximum delivery time to the processor.

The JR algorithm does not allow the processor to be idle if there is a ready job, even if the priority of the job is low. The IJR algorithm allows the processor to be idle while waiting for the more priority job. Additional conditions are checked (step b and c of IJR algorithm), under which it is profitable to perform the more priority job, but the downtime is not too large.

3   The Worst-Case Performance Ratio of the IJR Algorithm

The properties of the schedule created by algorithm IJR proposed in Sect. 2 are formulated and proved in the following lemmas.

Let the IJR algorithm constructs a schedule S, the value of the objective function is equal to C_A.

Consider some definitions that were introduced in [7] for schedules constructed according to Jackson's rule, and which are important characteristics of IIT schedules.

Definition 1. *[7] A critical job in the schedule S is a job c such that $C_A = \tau(c) + t(c) + q(c)$. If there are several such jobs, then we choose the earliest one (with minimum start time) in the schedule S.*

Definition 2. *[7] A critical sequence in a schedule S is a sequence of jobs $J(S) = (a, S^*, c)$ such that c is the critical job and there is no processor idle time in the schedule, starting from the start of the job a until the job c ends.*

The job a is either the first job in the schedule, or the processor is idle before it.

Definition 3. *[7] A job u in a critical sequence $J(S)$ is called interference job if $q(u) < q(c)$ and $q(i) \geq q(c)$, for all jobs i, that are done after job u in the critical sequence $J(S)$.*

Proposition 1. *[7] If for all jobs of the critical sequence it is true that $r(i) > r(a)$ and $q(i) \geq q(c)$, then the schedule is optimal.*

Let us introduce a definition of delayed job that can be encountered in IIT schedules.

Definition 4. *A job v from the critical sequence $J(S) = (a, S^*, c)$ is called a delayed job if $r(v) < r(a)$.*

An interference job can be a delayed job.

Let us formulate two properties of the schedule, like the properties of Jackson's rule schedules [7].

Lemma 1. *[18] Let there be the interference job u in the critical sequence $J(S) = (a, S_1, u, S_2, c)$. Let $r_{\min}(S_2) = \min\{r(i)|i \in S_2 \cup c\}$, $T(S_1) = \sum_{i \in S_1} t(i)$ and $idle = r_{\min}(S_2) - r(a) - t(a) - T(S_1) > 0$*
Then $C_A - C_{opt} \leq t(u) - idle$.

This lemma refines the property of the Jackson schedule, proved in [7], and shows that, removing the interference work, we can reduce the length of the schedule no more than $t(u) - idle$.

Lemma 2. *If there are not any delayed jobs in the critical sequence, then $C_A - C_{opt} \leq q(c)$.*

Lemma 3. *Let the IJR algorithm constructs a schedule S, the value of the objective function is equal to C_A.*
Assume the interference job u is in the critical sequence $J(S) = (a, S_1, u, S_2, c)$. If the job u is executed after the sequence S_2 and job c in an optimal schedule or the job u is performed between a and c in an optimal schedule then $C_A/C_{opt} \leq 3/2$.

Proof. Case 1. Let the job u is executed after the sequence S_2 and job c in an optimal schedule.
If $t(u) \leq C_{opt}/2$, then the Lemma 3 is true by Lemma 1.
Let $t(u) > C_{opt}/2$. Let $r_{\min}(S_2) = \min\{r(i)|i \in S_2 \cup c\}$. If the interference job u is executed after all jobs of the sequence S_2 and c in an optimal schedule, then

$C_{opt} \geq r_{\min}(S_2) + T(S_2) + t(c) + t(u) + q(u)$. Then

$$C_A - C_{opt} \leq r(a) + t(a) + T(S_1) + t(u) + T(S_2) + t(c)$$

$$+q(c) - r_{min}(S_2) - T(S_2) - t(c) - t(u) - q(u)$$

$$= r(a) + t(a) + T(S_1) - r_{min}(S_2) + q(c) - q(u)$$

$$= -idle + q(c) - q(u)$$

Choose a job $v \in S_2$ such that $r(v) = r_{\min}(S_2)$. The algorithm IJR did not put the job v instead of job u either because $q(v) < LB/2$ or $idle(v) > q(v) - q(u)$.

Then $idle = idle(v) = r_{\min}(S_2) - r(a) - t(a) - T(S_1) > q(v) - q(u) \geq q(c) - q(u)$, or $q(c) \leq q(v) < LB/2$.

Then $C_A - C_{opt} < LB/2$.

Case 2. If in the optimal schedule, the job u is performed between a and c, then .

$$C_{opt} \geq r(a) + t(a) + t(u) + T(S_2) + t(c) + q(c).$$

Therefore $C_A - C_{opt} \leq r(a) + t(a) + T(S_1) + t(u) + T(S_2) + t(c) + q(c) - r(a) - t(a) - t(u) - T(S_2) - t(c) - q(c) = T(S_1) \leq LB/2$. Because $t(u) > C_{opt}/2 \geq LB/2$ and $LB \geq t(a) + T(S_1) + t(u) + T(S_2) + t(c)$.

In this case, the IJR algorithm constructs a 3/2 approximation schedule.

Theorem 1. *The algorithm IJR constructs a schedule S_A for which $C_A/C_{opt} < 11/7$.*

Proof. Let the schedule S be constructed using the IJR algorithm. The objective function value is C_A, and there is the critical sequence $J(S) = (a, S^*, c)$ in S.

We consider all the possible cases.

Case 1. There are no interference and delayed jobs in $J(S)$ then the algorithm has constructed an optimal schedule.

Case 2. There are some delayed jobs in $J(S) = (a, S^*, c)$. If there is no interference job in the critical sequence $J(S)$, then $q(i) \geq q(c)$ for all jobs from the critical sequence $i \in J(S)$. But in the critical sequence there are jobs that can be started before the job a.

Let $r(J(S)) = \min\{r(i) \mid i \in J(S)\}$. Then

$$C_{opt} \geq r(J(S)) + T(J(S)) + q(c).$$

Hence

$$C_A - C_{opt} \leq r(a) + T(J(S)) + q(c) - r(J(S)) - T(J(S)) - q(c)$$

$$= r(a) - r(J(S)) < LB/2.$$

Because $q(a) > LB/2$.

Case 3. There is the interference job u in the critical sequence $J(S) = (a, S_1, u, S_2, c)$. It is required to consider the case in which $t(u) > C_{opt}/2$. If in an optimal schedule the job u is executed after the sequence S_2 and job c or the job u is performed between a and c, then $C_A/C_{opt} \leq 3/2$ by Lemma 3.

We have to consider the case, where in an optimal schedule the job u is executed before the job a.

Let us formulate a problem of fractional linear programming. We consider jobs that make up the critical sequence and introduce the following variables.

1. $x_1 = r(a)$ — release time of job a.
2. $x_2 = T(S_1)$ — sum of processing times jobs from S_1.
3. $x_3 = T(S_2)$ — sum of processing time jobs from S_2 and c.
4. $x_4 = t(u)$ — processing time of job u.
5. $x_5 = q(a)$ — delivery time of job a.
6. $x_6 = q(c)$ — delivery time of job c.
7. $x_7 = r_{\min}(S_2)$ — minimum release time of jobs from S_2.
8. $x_8 = C_{opt}$ — optimal value of objective function.
9. $x_9 = t(a)$ — processing time of job j_a.
10. $x_{10} = t(j)$ — maximum processing time of jobs from S_2 and c.
11. $x_{11} = idle$ — is a possible idle time of the processor if the algorithm puts a higher priority job, but not ready job, on the processor.

The objective function of the fractional linear programming problem is

$$C_A/C_{opt} = (x_1 + x_2 + x_3 + x_4 + x_6 + x_9)/x_8.$$

It is required to find its maximum value with the following restrictions:

1. The sum of the times of all jobs does not exceed the lower estimate:

$$x_2 + x_3 + x_4 + x_9 \leq LB.$$

2. For job a it is true $r(a) + t(a) + q(a) \leq LB$: $x_1 + x_5 + x_9 \leq LB$.
3. At time $x_1 + x_2 + x_9$ all tasks in the sequence S_2 are not ready and potential idle time of the processor is equal x_{11}

$$x_1 + x_2 - x_7 + x_9 + x_{11} \leq 0$$

4. For all jobs $i \in S_2$ it is true $r_{\min}(S_2) + t(i) + q(c) \leq LB$

$$x_6 + x_7 + x_{10} \leq LB$$

5. Optimal value of objective function $C_{opt} \geq r_{\min}(S_2) + T(S_2) + q(c)$.

$$x_3 + x_6 + x_7 - x_8 \leq 0$$

6. Optimal value of objective function $C_{opt} \geq t(u) + t(a) + q(a)$

$$x_4 + x_5 - x_8 + x_9 \leq 0$$

7. Job u jumped over a hence $q(a) \geq LB/2 + 1$: if LB is an even number and $q(a) \geq (LB+1)/2$ if LB is an odd number. $x_5 \geq LB/2+1$ or $x_5 \geq (LB+1)/2$.
8. Processing time of all jobs not less than 1: $x_9 \geq 1$, $x_{10} \geq 1$.
9. Idle time not less then 1: $x_{11} \geq 1$.

10. $x_i \geq 0, \forall i \in 1:11$.

Consider the case where LB is an even number, for odd LB the solution will be similar.

Let's change the variable $w = 1/x_8$ and $y_i = w * x_i$, then the conditions of the problem take the form:

$$F(\mathbf{y}) = y_1 + y_2 + y_3 + y_4 + y_6 + y_9 \longrightarrow \max$$

$$
\begin{aligned}
y_2 + y_3 + y_4 + && y_9 - && && LBw &\leq 0 \\
y_1 + && y_5 + && y_9 - && LBw &\leq 0 \\
y_1 + y_2 - && && y_7 + y_9 + \quad y_{11} && &\leq 0 \\
&& y_6 + y_7 + && y_{10} - && LBw &\leq 0 \\
y_3 + && y_6 + \quad y_7 && && &\leq 1 \\
y_4 + y_5 + && && y_9 && &\leq 1 \\
y_3 + y_4 + \quad y_6 + && && y_9 && &\leq 1 \\
-y_5 + && && && (LB/2+1)w &\leq 0 \\
&& -y_9 + && && w &\leq 0 \\
&& -y_{10} + && && w &\leq 0 \\
&& - && && y_{11} + w &\leq 0
\end{aligned}
$$

$$y_i \geq 0, i \in 1:11, w \geq 0$$

Let's construct a dual problem

$$g(\mathbf{u}) = u_5 + u_6 + u_7 \longrightarrow \min$$

$$
\begin{aligned}
u_2 + u_3 && && && &\geq 1 \\
u_1 + && u_5 + && u_7 && &\geq 1 \\
u_1 + && && u_6 + u_7 && &\geq 1 \\
u_2 + && u_6 - && u_8 && &\geq 0 \\
u_4 + u_5 + && u_7 && && &\geq 1 \\
-u_3 + \; u_4 + u_5 && && && &\geq 0 \\
u_1 + \quad u_2 + u_3 + && u_6 + u_7 - && u_9 && &\geq 1 \\
u_3 - && && && u_{11} &\geq 0 \\
u_4 - && && u_{10} && &\geq 0 \\
-LBu_1 - LB \quad u_2 - LBu_4 + && (LB/2+1)u_8 + u_9 + u_{10} + u_{11} && && &\geq 0
\end{aligned}
$$

$u_i \geq 0, i \in 1 : 11.$

Analyzing the third case, we could assert that all variables of the direct problem (except for x_2) must be different from zero. Hence, in the dual problem, the restrictions must be satisfied as equalities. Next, we solve this system of linear equations and obtain a solution to the dual problem.

Solving the dual problem:

$$u_1 = u_2 = u_4 = u_7 = u_{10} = (LB + 6)/(7LB + 2);$$

$$u_3 = u_8 = u_{11} = (6LB - 4)/(7LB + 2);$$

$$u_5 = u_6 = (5LB - 10)/(7LB + 2);$$

$$u_9 = 1.$$

$f(\mathbf{u}) = u_5 + u_6 + u_7 = 2(5LB - 10)/(7LB+2) + (LB+6)/(7LB+2) = (11LB - 14)/(7LB+2).$

Solving the primal problem:

1. $y_1 = (3LB - 12)(7LB + 2)$
2. $y_2 = (LB - 4)/(7LB + 2)$
3. $y_3 = (LB + 8)/(7LB + 2)$
4. $y_4 = (4LB - 10)/(7LB + 2)$
5. $y_5 = 3(LB + 2)/(7LB + 2)$
6. $y_6 = (2LB - 2)/(7LB + 2)$
7. $y_7 = (4LB - 4)/(7LB + 2)$
8. $w = 6/(7LB + 2)$
9. $y_9 = y_{10} = y_{11} = 6/(7LB + 2)$

$F(\mathbf{y}) = y_1 + y_2 + y_3 + y_4 + y_6 + y_9 = (3LB - 12)(7LB + 2) + (LB - 4)/(7LB + 2)$
$+ (LB + 8)/(7LB + 2) + (4LB - 10)/(7LB + 2) + (2LB - 2)/(7LB + 2) + 6/(7LB + 2) = (11LB - 14)/(7LB + 2).$

We have found feasible solutions to the primal and dual problems, and the values of the objective functions are equal. By the duality theorem, these solutions are optimal solutions direct and dual problems.

We can calculate x_j.

1. $x_1 = (3LB - 12)/6$
2. $x_2 = (LB - 4)/6$
3. $x_3 = (LB + 8)/6$
4. $x_4 = (4LB - 10)/6$
5. $x_5 = (LB/2 + 1)$
6. $x_6 = (LB - 1)/3$
7. $x_7 = 2(LB - 1)/3$
8. $x_8 = (7LB + 2)/6$
9. $x_9 = x_{10} = x_{11} = 1$

$f(\mathbf{x}) = (11LB - 14)/(7LB + 2)$

Solving the similar problem of fractional linear programming for the case of an odd LB, we obtain the following solution.

Table 2. Release, processing and delivery times of jobs

Job	a	u	S_1	j_1	...	j_{17}	c
r_i	48	0	48	65	65	65	65
t_i	1	65	16	1	1	1	1
q_i	51	0	1	33	33	33	33

Table 3. IJR schedule

48	1	16	65	1	16	1	1	33
idle	a	S_1	u	j_1	...	j_{17}	c	q(c)

1. $x_1 = (LB - 3)/2$
2. $x_2 = (LB - 5)/6$
3. $x_3 = (LB + 7)/6$
4. $x_4 = (2LB - 4)/3$
5. $x_5 = (LB + 1)/2$
6. $x_6 = (LB - 2)/3$
7. $x_7 = (2LB - 1)/3$
8. $x_8 = (7LB + 1)/6$
9. $x_9 = x_{10} = x_{11} = 1$

$f(\mathbf{x}) = (11LB - 13)/(7LB + 1)$

We have proven that the worst-case performance ratio of IJR algorithm is 11/7.

The computational complexity of the IJR algorithm was established in [17].

Lemma 4. *[17] The computational complexity of the IJR algorithm is* $O(n \log n)$.

Example 1. Consider a system of 21 jobs : a, u, S_1, c and 17 identical tasks, which are included in S_2. The data for the job system are given in Table 2.

The lower bound for the objective function is equal $LB = 100$.
The IJR algorithm constructs the schedule $S = (a, S_1, u, j_1, \ldots, j_{17}, c)$ (see Table 3). The processor is idle 48 time units before starting the job a. Job c is the critical job, the execution of job c ends at 148 and the delivery of job c ends at 181.

The objective function is equal $C_{\max}(S) = LB/2 - 2 + 1 + (LB - 4)/6 + (LB + 8)/6 + (2LB - 5)/3 + (LB - 1)/3 = (11LB - 14)/6 = 181$.

The optimal schedule is $S_{opt} = (u, a, S_2, c, S_1)$ (see Table 4), the value of the objective function for which is equal $C_{opt} = 117$.

$C_A/C_{opt} = (11LB - 14)(7LB + 2) = 1.547$.

We can build a similar example for odd value $LB = 101$.

Table 4. Optimal schedule

65	1	17	1	16
u	a	S_2	c	S_1

4   Computational Experiment

To find out the practical efficiency of the algorithm, a computational experiment was carried out. The goal of the computational experiment was to compare of the accuracy of the IJR algorithm with the accuracy of the JR Schrage algorithm and with the accuracy of the combined ICA (Idle Combined Algorithm) algorithm, in which the best solution was chosen of the two solutions, obtained by the JR and IJR algorithms.

The initial data was generated by the method described by Carlier [11]. For each task $i \in 1 : n$, two integer values were chosen with uniform distribution: $q(i)$ between 1 and nK, $r(i)$ between 1 and nK. There were chosen $n = 100$ and the values for K from 14 to 18, the ranges, which hard instances occurred most frequently.[?]. For each value of K, we considered 200 instances. Three groups of examples were considered. The processing times $t(j)$ of jobs from each group were selected from the following intervals ($t_{\max} = 50$):

1. Type A: $t(j)$ from $[1, t_{\max}]$,
2. Type B: $t(j)$ from $[1, t_{\max}/2]$, for $j \in 1 : n - 1$ and
   $t(j_n)$ from $[nt_{\max}/8, 3nt_{\max}/8]$,
3. Type C: $t(j)$ from $[1, t_{\max}/3]$, for $j \in 1 : n - 2$ and $t(j_{n-1}), t(j_n)$ from $[nt_{\max}/12, 3nt_{\max}/12]$.

Type B groups contains instances with one long job and type C groups contains instances with two long jobs.

The value of the objective function C_A was compared with the optimal value of the objective function C_{opt}, which was obtained by the branch and bound method [18]. In all tables, n is the number of tasks in the instance. In the second column there is constant K, average relative error $R_{IJR} = C_{IJR}/C_{opt}$ for IJR algorithm in the 3 column, average relative error $R_{JR} = C_{JR}/C_{opt}$ for JR algorithm in the 4 column, and average relative error $R_{ICA} = C_{ICA}/C_{opt}$ for ICA algorithm in the 5 column.

For instances of Type A the average relative error of the solution is small (at most 0.97 %) for all algorithms and decreases with increasing n. We established that changing the constant K from 10 to 22 does not significantly affect the results of the algorithms for instances of Type A.

The theoretical analysis of the algorithms shows that the most difficult examples take place when there are one or two long tasks. Such tests were generated in groups of type B and type C.

The value of the average relative error of algorithms for tests of type B are given in the Table 5.

Table 5. Type B. The average relative error of algorithms.

n	K	R_{IJR}	R_{JR}	R_{ICA}
100	14	1.06	1.03	1.004
100	15	1.05	1.04	1.007
100	16	1.01	1.04	1.004
100	17	1.02	1.03	1.006
100	18	1.05	1.06	1.005

Table 6. Type C. The average relative error of algorithms.

n	K	R_{IJR}	R_{JR}	R_{ICA}
100	14	1.02	1.04	1.004
100	15	1.07	1.03	1.006
100	16	1.05	1.04	1.005
100	17	1.06	1.04	1.008
100	18	1.06	1.05	1.004

The relative error of the solution increases for algorithms JR and IJR it is from 1 to 6 % on average. The ICA algorithm has significantly more advantages. It combines the advantages of the Schrage algorithm, which does not allow unforced idle time and IJR algorithm, which allows them. The relative error of the solution for ICA algorithm is from 1.004 to 1.007 on average.

Tables 6 show the results of comparison of algorithms for tests of Type C.

The combined ICA algorithm significantly diminishes the relative error of the solution. For the combined algorithm it ranges from 0.4 to 0.8 %.

5   Conclusion

The paper considers the problem of scheduling for single processor with release and delivery times. The goal is minimization of the total execution time of all jobs. The paper proposes the IJR algorithm, in which the priority of the job is considered first and processor can be idle, when certain conditions are met. IJR algorithm builds only one permutation and its computational complexity is $O(n \log n)$.

We prove that the worst-case performance ratio of the algorithm IJR is equal 11/7. The computational experiment has confirmed the practical efficiency of the IJR algorithm.

References

1. Graham, R.L., Lawner, E.L., Rinnoy Kan, A.H.G.: Optimization and approximation in deterministic sequencing and scheduling. A survey. Ann. Disc. Math. **5**(10), 287–326 (1979)

2. Lenstra, J.K., Rinnooy Kan, A.H.G., Brucker, P.: Complexity of machine scheduling problems. Ann. Disc. Math. **1**, 343–362 (1977)
3. Artigues, C., Feillet, D.: A branch and bound method for the job-shop problem with sequence-dependent setup times. Ann. Oper. Res. **159**, 135–159 (2008)
4. Chandra, C., Liu, Z., He, J., Ruohonen, T.: A binary branch and bound algorithm to minimize maximum scheduling cost. Omega **42**, 9–15 (2014)
5. Sourirajan, K., Uzsoy, R.: Hybrid decomposition heuristics for solving large-scale scheduling problems in semiconductor wafer fabrication. J. Sched. **10**, 41–65 (2007)
6. Schrage, L.: Optimal solutions to resource constrained network scheduling problems (unpublished manuscript) (1971)
7. Potts, C.N.: Analysis of a heuristic for one machine sequencing with release dates and delivery times. Oper. Res. Int. Journal **28**(6), 445–462 (1980)
8. Hall, L.A., Shmoys, D.B.: Jackson's rule for single-machine scheduling: making a good heuristic better. Math. Oper. Res. **17**(1), 22–35 (1992)
9. Nowicki, E., Smutnicki, C.: An approximation algorithm for a single-machine scheduling problem with release times and delivery times. Discrete Appl. Math. **48**, 69–79 (1994)
10. Baker, K.R.: Introduction to Sequencing and Scheduling. Wiley, New York (1974)
11. Carlier, J.: The one machine sequencing problem. Eur. J. Oper. Res. **11**, 42–47 (1982)
12. Grabowski, J., Nowicki, E., Zdrzalka, S.: A block approach for single-machine scheduling with release dates and due dates. Eur. J. Oper. Res. **26**, 278–285 (1986)
13. McMahon, G.B., Florian, N.: On scheduling with ready times and due dates to minimize maximum lateness. Oper. Res. **23**(3), 475–482 (1975)
14. Pan, Y., Shi, L.: Branch and bound algorithm for solving hard instances of the one-machine sequencing problem. Eur. J. Oper. Res. **168**, 1030–1039 (2006)
15. Liu, Z.: Single machine scheduling to minimize maximum lateness subject to release dates and precedence constraints. Comput. Oper. Res. **37**, 1537–1543 (2010)
16. Kanet, J., Sridharan, V.: Scheduling with inserted idle time: problem taxonomy and literature review. Oper. Res. **48**(1), 99–110 (2000)
17. Grigoreva, N.: Worst-case analysis of an approximation algorithm for single machine scheduling problem. In: Proceedings of the 16th Conference on Computer Science and Intelligence Systems, vol. 25, pp. 221-225. (Annals of Computer Science and Information System; v. 25) (2021)
18. Grigoreva, N.: Single machine scheduling with precedence constrains, release and delivery times. In: Wilimowska, Z., Borzemski, L., Świątek, J. (eds.) ISAT 2019. AISC, vol. 1052, pp. 188–198. Springer, Cham (2020). https://doi.org/10.1007/978-3-030-30443-0_17

Optimization and Data Analysis

Decentralized Strongly-Convex Optimization with Affine Constraints: Primal and Dual Approaches

Alexander Rogozin[1(✉)], Demyan Yarmoshik[1], Ksenia Kopylova[2],
and Alexander Gasnikov[1,3,4]

[1] Moscow Institute of Physics and Technology, Moscow, Russia
aleksandr.rogozin@phystech.edu
[2] Saint Petersburg State University, Saint Petersburg, Russia
[3] Caucasus Mathematical Center, Adyghe State University, Maikop, Russia
[4] IITP RAS, Moscow, Russia

Abstract. Decentralized optimization is a common paradigm used in distributed signal processing and sensing as well as privacy-preserving and large-scale machine learning. It is assumed that several computational entities locally hold objective functions and are connected by a network. The agents aim to commonly minimize the sum of the local objectives subject by making gradient updates and exchanging information with their immediate neighbors. Theory of decentralized optimization is pretty well-developed in the literature. In particular, it includes lower bounds and optimal algorithms. In this paper, we assume that along with an objective, each node also holds affine constraints. We discuss several primal and dual approaches to decentralized optimization problem with affine constraints.

Keywords: Distributed optimization · Convex optimization · Constrained optimization

1 Introduction

Many distributed systems such as distributed sensor networks, systems for power flow control and large-scale architectures for machine learning use decentralized optimization as a basic mathematical tool. Several applications such as power systems control [11,17] lead to problems where the agents locally hold optimization objectives and aim to cooperatively minimize the sum of the objectives. Moreover, every node locally holds affine constraints for its decision variable.

The work of D. Yarmoshik in Sects. 1, 6, 7 was supported by the program "Leading Scientific Schools" (grant no. NSh-775.2022.1.1). The work of A. Rogozin and A. Gasnikov in Sects. 2–5 was supported by Russian Science Foundation (project No. 21-71-30005).

N. Olenev et al. (Eds.): OPTIMA 2022, CCIS 1739, pp. 93–105, 2022.
https://doi.org/10.1007/978-3-031-22990-9_7

Decentralized optimization without affine constraints can be called a well-examined area of research. It is known that the performance of optimization algorithms executed over strongly-convex smooth objectives is lower bounded by a multiple of the graph condition number and objective condition number (up to a logarithmic factor) [19]. Both primal [8] and dual [19] algorithms that reach the lower bounds have been proposed. The algorithms are based on reformulating network communication constraints as affine constraints via a communication matrix associated with the network (i.e. Laplacian matrix). Introduction of affine constraints at the nodes leads to new classes of algorithms that can be divided into two main types. The first type are consensus-based methods that can be either primal or dual [2,9,10,12–14,23]. The second type are ADMM-based methods [1,3,18,21]. Let us briefly review some of the closely related papers.

The paper [12] is dedicated to constrained distributed optimization and consider only separable objective functions (each agent has its own independent variable). Moreover, affine constraints are supposed to be network-compatiable (constraint matrix can have a non-zero element on position (i,j) only if there is an edge in communication graph between agents i and j). We do not impose such limitations: in our case each term in the objective functions depends on the same shared variable (formulation in [12] is obviously a special case of this) and matrix of constraints can have arbitrary structure.

In [14] the authors present various formulations of distributed optimization problems with different types of interconnections between constraints and objectives, including the case, when the objective (cost) cannot be represented as sum of cost functions of each agent. However, their algorithms for problems with coupled affine constraints require to solve a "master problem" on central node at each iteration and thus are not decentralized.

The authors of [20] consider multi-cluster distributed problem formulation which is a generalization of multi-agent approach. In multi-cluster case agents within one cluster have the same decision variable while different clusters corresponds to different decision variables. All variables are subject to a coupled affine constraint. By incorporating consensus constraints into dual problem with Lagrangian multipliers the author comes to solving a saddle point problem and prove asymptotic $O(1/N)$ ergodic convergence rate for their method. Dependency of convergence rate on problem parameters in saddle point approach was studied in [22].

Our paper studies the application of different techniques to decentralized problems with affine constraints. We obtain linear convergence rates with (explicitly specified) accelerated dependencies on function properties, constraint matrix spectrum and communication graph properties.

The paper outline is as follows. In Sect. 4 we discuss a primal approach, that is based on reformulation the initial distributed problem as a saddle-point problem and applying algorithm of paper [7] afterwards. In Sect. 5, we describe a method that allows to incorporate both affine and communication constraints to the dual function. We refer the approach in Sect. 5 as a globally dual approach. Finally, in Sect. 6 we describe a slightly different dual approach that firstly takes dual functions locally at the nodes and incorporates consensus constraints afterwards. We refer to the latter method as a locally dual approach.

2 Preliminaries

Let $\mathrm{col}(x_1, \ldots, x_m)$ define a column vector of $x_1, \ldots, x_m \in \mathbb{R}^d$, i.e. $\mathrm{col}(x_1, \ldots, x_m) = [x_1^\top \ldots x_m^\top]^\top$. For matrices P and Q, their Kronecker product is defined as $P \otimes Q$. Identity matrix of size $p \times p$ is denoted $\mathbf{I}_p$. Moreover, given a symmetric positive semi-definite matrix, we denote $\lambda_{\max}(\cdot)$, $\lambda_{\min}(\cdot)$, $\lambda_{\min}^+(\cdot)$ its maximal, minimal and minimal nonzero eigenvalues, respectively. We also let $\sigma_{\max}(\cdot)$, $\sigma_{\min}(\cdot)$ and $\sigma_{\min}^+(\cdot)$ be the maximal, minimal and minimal nonzero singular values of a matrix, respectively.

In the forthcoming analysis, we will need the following basic lemma concerning Kronecker product properties.

Lemma 1. *Given two matrices P and Q such that $\sigma_{\min}(P) = \sigma_{\min}(Q) = 0$, we have*

$$\sigma_{\max}(P \otimes \mathbf{I} + \mathbf{I} \otimes Q) = \sigma_{\max}(P) + \sigma_{\max}(Q),$$
$$\sigma_{\min}^+(P \otimes \mathbf{I} + \mathbf{I} \otimes Q) = \min\left\{\sigma_{\min}^+(P), \sigma_{\min}^+(Q)\right\}$$

Proof. Consider decompositions $P = U_P \Sigma_P V_P^\top$ and $Q = U_Q \Sigma_Q V_Q^\top$, where U_P, V_P, U_Q, V_Q are orthogonal matrices and Σ_P and Σ_Q are diagonal matrices with corresponding eigenvalues at the diagonal. We have

$$(U_P^\top \otimes U_Q^\top)(P \otimes \mathbf{I} + \mathbf{I} \otimes Q)(V_P \otimes V_Q) = \Sigma_P \otimes \mathbf{I} + \mathbf{I} \otimes \Sigma_Q.$$

Denote singular values of P as $\alpha_1, \ldots, \alpha_n$ and the singular values of Q as $\beta_1, \ldots, \beta_m$. Singular values of $P \otimes \mathbf{I} + \mathbf{I} \otimes Q$ have form

$$\lambda(\alpha_i, \beta_j) = \alpha_i + \beta_j, \ i = 1, \ldots, n, \ j = 1, \ldots, m.$$

Therefore, $\sigma_{\max}(P \otimes \mathbf{I} + \mathbf{I} \otimes Q) = \sigma_{\max}(P) + \sigma_{\max}(Q)$. For the minimal nonzero singular values we obtain

$$\sigma_{\min}^+(P \otimes \mathbf{I} + \mathbf{I} \otimes Q) = \min\left\{\sigma_{\min}^+(P), \sigma_{\min}^+(Q)\right\}.$$

3 Problem Statement

Consider minimization problem with affine constraints.

$$\min_{x \in \mathbb{R}^d} \sum_{i=1}^m f_i(x) \ \text{s.t.} \ Bx = 0. \tag{1}$$

We assume that each f_i is held by a separate agent, and the agents can exchange information through some communication network. Each agent also locally holds affine optimization constraints $Bx = 0$, where $B \in \mathbb{R}^{p \times d}$. Further we assume that Ker $B \neq \{0\}$, because otherwise the constraints $Bx = 0$ define a set consisting of only $\{0\}$, which is not an interesting case.

We make assumptions on the optimization objectives that are standard for optimization literature [16].

Assumption 1. *Each f_i ($i = 1, \ldots, m$) is differentiable, μ-strongly convex and L-smooth, i.e.*

$$f(y) \geq f(x) + \langle \nabla f(x), y - x \rangle + \frac{\mu}{2} \|y - x\|_2^2,$$

$$f(y) \leq f(x) + \langle \nabla f(x), y - x \rangle + \frac{L}{2} \|y - x\|_2^2.$$

The communication network is represented by an undirected connected graph $\mathcal{G} = (\mathcal{V}, \mathcal{E})$. The communication constraints are represented by a specific matrix W associated with the graph $\mathcal{G}$.

Assumption 2

1. W *is a symmetric positive semi-definite matrix.*
2. *(Network compatibility) For all $i, j = 1, \ldots, m$ it holds $[W]_{ij} = 0$ if $(i, j) \notin \mathcal{E}$ and $i \neq j$.*
3. *(Kernel property) For any $v = [v_1, \ldots, v_m]^\top \in \mathbb{R}^m$, $Wv = 0$ if and only if $v_1 = \ldots = v_m$, i.e. Ker $W = \operatorname{span}\{\mathbf{1}\}$.*

An explicit example of a matrix that satisfies Assumption 2 is the Graph Laplacian $W \in \mathbb{R}^{m \times m}$:

$$[W]_{ij} \triangleq \begin{cases} -1, & \text{if } (i, j) \in E, \\ \deg(i), & \text{if } i = j, \\ 0, & \text{otherwise.} \end{cases} \tag{2}$$

Let us introduce $\mathbf{x} = \operatorname{col}(x_1 \ldots x_m)$ and $\mathbf{W} = W \otimes \mathbf{I}$. According to Assumption 2, communication constraints $x_1 = \ldots = x_m$ can be equivalently rewritten as $\mathbf{W}\mathbf{x} = 0$. Also introduce $\mathbf{B} = \mathbf{I} \otimes B$ and $F(\mathbf{x}) = \sum_{i=1}^m f_i(x_i)$. That allows to rewrite problem (1) in the following way.

$$\min_{\mathbf{x} \in \mathbb{R}^{md}} F(\mathbf{x}) \tag{3}$$

$$\text{s.t. } \mathbf{W}\mathbf{x} = 0, \ \mathbf{B}\mathbf{x} = 0.$$

Reformulation 3 admits implementation of optimization methods for affinely constrained minimization. The iterations of such methods become automatically decentralized in the following sense. Let the optimization algorithm use primal or dual oracle calls of the objective function and use multiplications by the matrices representing affine constraints. In the case of problem (3) the gradient $\nabla F(\mathbf{x}) = \operatorname{col}[\nabla f_1(x_1) \ldots \nabla f_m(x_m)]$ is computed locally on the nodes and stored in a distributed manner across the network. Multiplication by $\mathbf{B}$ is also performed locally due to its definition (i.e. the i-th node computes Bx_i), and the multiplication by $\mathbf{W}$ is performed in a decentralized manner due to the network compatibility property of W (see Assumption 2).

4   Primal Approach

In this section, we discuss the solution of problem (3) by an algorithm APDG [7] that only uses primal oracle calls. The algorithm is designed for saddle-point problems, so we reformulate (3) as a saddle-point problem.

We add dual multipliers for the constraints and get a saddle-point problem

$$\min_{\mathbf{x}\in\mathbb{R}^{md}} \max_{\mathbf{u}\in\mathbb{R}^{mp},\mathbf{v}\in\mathbb{R}^{md}} F(\mathbf{x}) + \langle \mathbf{u}, \mathbf{Bx}\rangle + \gamma\langle \mathbf{v}, \mathbf{Wx}\rangle = F(\mathbf{x}) + \left\langle \begin{pmatrix} \mathbf{u} \\ \mathbf{v} \end{pmatrix}, \begin{pmatrix} \mathbf{B} \\ \gamma\mathbf{W} \end{pmatrix}\mathbf{x} \right\rangle.$$

$$(4)$$

Algorithm 1. APDG: Accelerated Primal-Dual Gradient Method

1: **Input:** $\mathbf{x}^0 \in \operatorname{Range}\mathbf{A}^\top, \mathbf{y}^0 \in \operatorname{Range}\mathbf{A}, \eta_x, \eta_y, \alpha_x, \beta_x, \beta_y > 0, \tau_x, \tau_y, \sigma_x, \sigma_y \in (0,1],$
   $\theta \in (0,1)$
2: $\mathbf{x}_f^0 = \mathbf{x}^0$
3: $\mathbf{y}_f^0 = \mathbf{y}^{-1} = \mathbf{y}^0$
4: **for** $k = 0, 1, 2, \ldots$ **do**
5:   $\mathbf{y}_m^k = \mathbf{y}^k + \theta(\mathbf{y}^k - \mathbf{y}^{k-1})$
6:   $\mathbf{x}_g^k = \tau_x\mathbf{x}^k + (1-\tau_x)\mathbf{x}_f^k$
7:   $\mathbf{y}_g^k = \tau_y\mathbf{y}^k + (1-\tau_y)\mathbf{y}_f^k$
8:   $\mathbf{x}^{k+1} = \mathbf{x}^k + \eta_x\alpha_x(\mathbf{x}_g^k - \mathbf{x}^k) - \eta_x\beta_x\mathbf{A}^\top\mathbf{A}\mathbf{x}^k - \eta_x\left(\nabla F(\mathbf{x}_g^k) + \mathbf{A}^\top\mathbf{y}_m^k\right)$
9:   $\mathbf{y}^{k+1} = \mathbf{y}^k - \eta_y\beta_y\mathbf{A}(\mathbf{A}^\top\mathbf{y}^k + \nabla F(\mathbf{x}_g^k)) + \eta_y\mathbf{A}\mathbf{x}^{k+1}$
10: $\mathbf{x}_f^{k+1} = \mathbf{x}_g^k + \sigma_x(\mathbf{x}^{k+1} - \mathbf{x}^k)$
11: $\mathbf{y}_f^{k+1} = \mathbf{y}_g^k + \sigma_y(\mathbf{y}^{k+1} - \mathbf{y}^k)$
12: **end for**

Denote $\mathbf{A} = \begin{pmatrix} \mathbf{B} \\ \gamma\mathbf{W} \end{pmatrix}$. In order to get complexity bounds for APDG applied to problem (4), we need to bound the spectrum of $\mathbf{A}$. Note that $\mathbf{A}^\top\mathbf{A} = \mathbf{B}^\top\mathbf{B} + \gamma^2\mathbf{W}^2 = \mathbf{I}_m \otimes (B^\top B) + \gamma^2 W^2 \otimes \mathbf{I}_d$. By Lemma 1 we have

$$\lambda_{\max}(\mathbf{A}^\top\mathbf{A}) = \lambda_{\max}(B^\top B) + \gamma^2\lambda_{\max}^2(W),$$
$$\lambda_{\min}^+(\mathbf{A}^\top\mathbf{A}) = \min\left\{\lambda_{\min}^+(B^\top B), \gamma^2(\lambda_{\min}^+(W))^2\right\}.$$

We can also compute the condition number of $\mathbf{A}^\top\mathbf{A}$:

$$\chi(\mathbf{A}^\top\mathbf{A}) = \frac{\lambda_{\max}(\mathbf{A}^\top\mathbf{A})}{\lambda_{\min}^+(\mathbf{A}^\top\mathbf{A})} = \frac{\lambda_{\max}(B^\top B) + \gamma^2\lambda_{\max}^2(W)}{\min\left\{\lambda_{\min}^+(B^\top B), \gamma^2(\lambda_{\min}^+(W))^2\right\}}.$$

By accurately choosing factor γ, we can control the condition number $\chi(\mathbf{A}^\top \mathbf{A})$. The minimal value of $\chi(\mathbf{A}^\top \mathbf{A})$ is attained at $\gamma^2 = \frac{\lambda_{\min}^+(B^\top B)}{(\lambda_{\min}^+(W))^2}$ and equals $\chi(\mathbf{A}^\top \mathbf{A}) = \chi(B^\top B) + \chi^2(W)$. Therefore, if we apply APDG directly to problem (3), the complexity would be

$$O\left(\max\left(\sqrt{\chi^2(W) + \chi(B^\top B)}\sqrt{\frac{L}{\mu}}, \chi^2(W) + \chi(B^\top B)\right)\log\frac{1}{\varepsilon_{\mathbf{x}}}\right)$$

calls of $\nabla f_i(\cdot)$ at each node and communication rounds, with $\varepsilon_{\mathbf{x}}$ being the desired distance to the solution: $\|\mathbf{x}^N - \mathbf{x}^*\| \leq \varepsilon_{\mathbf{x}}$. In the smooth, strongly convex case it is also the complexity for satisfying $F(\mathbf{x}^N) - F(\mathbf{x}) \leq \varepsilon_F$ or $\|\mathbf{A}\mathbf{x}^N\| \leq \varepsilon_{\mathbf{A}}$ (up to logarithmic dependencies on the problem parameters). Indeed, from Lipschitz smoothness we have $F(\mathbf{x}^N) - F(\mathbf{x}) \leq L\varepsilon_{\mathbf{x}}^2/2$ and $\|\mathbf{A}\mathbf{x}^N\| = \|\mathbf{A}\mathbf{x}^N - \mathbf{A}\mathbf{x}^*\| \leq \sigma_{max}(\mathbf{A})\varepsilon_{\mathbf{x}}$. By that means, in the following inequalities ε can be replaced by any of $\varepsilon_{\mathbf{x}}$, ε_f, $\varepsilon_{\mathbf{A}}$.

The dependence on network parameters W and affine constraints parameters B can be enhanced by using Chebyshev acceleration [19]. Let us replace W by a Chebyshev polynomial $P_K(W)$ such that it has degree $K = O\left(\sqrt{\chi(W)}\right)$ and condition number $\chi(P_K(W)) = O(1)$. Multiplication by $P_K(W)$ is equivalent to making K communication rounds. Analogically, let us replace $B^\top B$ by a Chebyshev polynomial $P_M(B^\top B)$ with degree $M = O\left(\sqrt{\chi(B^\top B)}\right)$ and condition number $\chi\left(P_M(B^\top B)\right) = O(1)$. As a result, we obtain

$$N = O\left(\sqrt{\frac{L}{\mu}}\log\frac{1}{\varepsilon}\right) \text{ oracle calls at each node,}$$

$$O\left(N\sqrt{\chi(W)}\right) \text{ communications,}$$

$$O\left(N\sqrt{\chi(B^\top B)}\right) \text{ multiplications by } B, B^\top \text{ at each node.}$$

5   Globally Dual Approach

In this section, we describe an approach to solving (3) that is based on passing to the dual problem. We call this approach "global" since both constraints, that is, affine constraints $\mathbf{B}\mathbf{x} = 0$ and communication constraints $\mathbf{W}\mathbf{x} = 0$ are used in the dual reformulation.

Let γ be a positive scalar and $\mathbf{A}^\top = [\mathbf{B}^\top \ \gamma\mathbf{W}]$ and introduce dual function

$$\Phi(\mathbf{u}) = \max_{\mathbf{x}\in\mathbb{R}^{md}}[-F(\mathbf{x}) + \langle\mathbf{u}, \mathbf{A}\mathbf{x}\rangle] = F^*(\mathbf{A}^\top\mathbf{u}).$$

We have $\nabla\Phi(\mathbf{u}) = \mathbf{A}\nabla F^*(\mathbf{A}^\top\mathbf{u}) = \mathbf{A} \cdot \arg\min_{\mathbf{x}\in\mathbb{R}^{md}}[-F(\mathbf{x}) + \langle\mathbf{u}, \mathbf{A}\mathbf{x}\rangle]$. Note that multiplication by $\mathbf{A}$ is performed in a distributed manner: indeed, it includes local multiplications by B and a consensus round, which is a multiplication by $\mathbf{W}$. Moreover, the arg min operation is computed locally, which is standard for decentralized optimization [19]. Finally, dual function Φ is $\frac{\lambda_{\max}(\mathbf{A}^\top\mathbf{A})}{\mu}$-smooth on $\mathbb{R}^{2md}$ and $L_\Phi = \frac{\lambda^+_{\min}(\mathbf{A}^\top\mathbf{A})}{L}$-strongly convex on $(\operatorname{Ker}\mathbf{A}^\top)^\perp$. Solving dual problem

$$\min_{\mathbf{u}\in\mathbb{R}^{2md}} \Phi(\mathbf{u})$$

by a fast gradient method (see i.e. accelerated Nesterov method in Sect. 2.2 of [16]) until accuracy $\Phi(\mathbf{u}^N) - \Phi(\mathbf{u}) \le \varepsilon_\Phi$ requires $N = O\left(\sqrt{\frac{L}{\mu}}\sqrt{\chi(\mathbf{A}^\top\mathbf{A})}\log\frac{1}{\varepsilon_\Phi}\right)$ iterations.

Following the same arguments as in Sect. 4, we compute the condition number $\chi(\mathbf{A}^\top\mathbf{A})$:

$$\chi(\mathbf{A}^\top\mathbf{A}) = \frac{\lambda_{\max}(B^\top B) + \gamma^2\lambda^2_{\max}(W)}{\min\left\{\lambda^+_{\min}(B^\top B), \gamma^2(\lambda^+_{\min}(W)^2)\right\}}.$$

The minimal value of $\chi(\mathbf{A}^\top\mathbf{A})$ is attained at $\gamma^2 = \frac{\lambda^+_{\min}(B^\top B)}{(\lambda^+_{\min}(W))^2}$ and equals $\chi(\mathbf{A}^\top\mathbf{A}) = \chi(B^\top B) + \chi^2(W)$. Communication and computation complexities of fast dual method equal

$$O\left(\sqrt{\frac{L}{\mu}}\left(\chi(B^\top B) + \chi^2(W)\right)^{\frac{1}{2}}\log\frac{1}{\varepsilon_\Phi}\right).$$

To obtain desired complexity estimates for the algorithm to find the approximate solution $\mathbf{x}^N$ satisfying $F(\mathbf{x}^N) - F(\mathbf{x}) \le \varepsilon$ and $\|\mathbf{A}\mathbf{x}^N\| \le \varepsilon$, we refer to the following properties of dual function (see, e.g. Theorem 5.2 from [4]):

$$\|\nabla\Phi(\mathbf{u})\| \le \epsilon/R_\mathbf{u} \Rightarrow F(\mathbf{x}(\mathbf{u})) - F(\mathbf{x}^*) \le \epsilon,$$
$$\|\nabla\Phi(\mathbf{u})\| \le \epsilon \Rightarrow \|\mathbf{A}\mathbf{x}(\mathbf{u})\| < \epsilon,$$

where $\|\mathbf{u}\| \le 2R_\mathbf{u}$, and $\mathbf{x}(\mathbf{u}) = \arg\min_{\mathbf{x}\in\mathbb{R}^{md}} [-F(\mathbf{x}) + \langle\mathbf{u}, \mathbf{A}\mathbf{x}\rangle]$. Combining it with $\Phi(\mathbf{u}^N) - \Phi(\mathbf{u}) \ge \|\Phi(\mathbf{u}^N)\|^2/2L_\Phi$, which is true for a smooth convex function, we justify substitution of ε_Φ by ε in the complexity estimate. This transition will only change the constant hidden by big-O notation (by the factor of two), and affect omitted logarithmic dependencies on the problem parameters.

To employ Chebyshev acceleration in this case we do substitution $\mathbf{A}^\top\mathbf{u} \to \mathbf{p}$. In this variables accelerated Nesterov method turns into Algorithm 2, where $\mathbf{x}(\mathbf{q}) = \nabla F^*(\mathbf{q}) = \arg\min[-F(\mathbf{x}) + \langle\mathbf{q}, x\rangle]$:

Algorithm 2. Globally Dual Method

1: **Input:** $\mathbf{p}^0 \in \operatorname{Range}\mathbf{A}^\top$, $\eta > 0$, $\beta \in (0, 1)$
2: $\mathbf{p}^{-1} = \mathbf{p}^0$
3: **for** $k = 0, 1, 2, \ldots$ **do**
4: $\quad \mathbf{q} = \mathbf{p}^k + \beta\left(\mathbf{p}^k - \mathbf{p}^{k-1}\right)$
5: $\quad \mathbf{p}^{k+1} = \mathbf{q} - \eta\mathbf{A}^\top\mathbf{A}\mathbf{x}(\mathbf{q})$
6: **end for**

For the algorithm in this form we can replace $\mathbf{A}^\top\mathbf{A}$ with Chebyshev polynomial of it, as we did in Sect. 4, and obtain the same complexity estimates as for APDG:

$$N = O\left(\sqrt{\frac{L}{\mu}}\log\frac{1}{\varepsilon}\right) \text{ oracle calls at each node,}$$

$$O\left(N\sqrt{\chi(W)}\right) \text{ communications,}$$

$$O\left(N\sqrt{\chi(B^\top B)}\right) \text{ multiplications by } B, B^\top \text{ at each node.}$$

6   Locally Dual Approach

In Sect. 5 we discussed a dual reformulation of (3) where both constraints $\mathbf{Bx} = 0$ and $\mathbf{Wx} = 0$ are used simultaneously. This section describes a dual approach, as well, but the difference is that we firstly pass to dual functions locally at the nodes and impose the communication constraints only afterwards.

6.1   Utilizing Locality on y

One can note that in the above approaches optimization over $\mathbf{y}$ could be done locally at each node. This is equivalent to including affine constraints into the objective (as an indicator function) instead of handling them with Lagrangian multipliers. In settings there the "cost" of communication is limiting or comparable to that of local computations, we can find the solution faster by going this way. It may be the case when x has a small dimension and decentralization is desirable due to privacy constraints.

Dual problem in this approach will be

$$\max_{\mathbf{z}} \min_{\mathbf{Bx}=0} \{F(\mathbf{x}) + \langle\mathbf{z}, \mathbf{Wx}\rangle\} = -\min_{\mathbf{z}} F^*_{[\mathbf{Bx}=0]}(\mathbf{W}^\top\mathbf{z}),$$

where $F^*_{[\mathbf{Bx}=0]}(\mathbf{z}) = \max_{\mathbf{Bx}=0}\{\langle\mathbf{z}, \mathbf{x}\rangle - F(\mathbf{x})\}$ denotes a convex conjugate under affine constraints.

We can reduce the problem of computing the gradient of such a modified conjugate function to calling conventional dual oracle. Let E be a matrix, the rows of which constitute an orthogonal basis in the null space of B (matrix E can be computed at the preprocessing stage of an optimization algorithm). Then instead of working with functions $f_i(x)$ we can optimize the sum of functions $h_i(t) = f_i(Et)$.

Denote $\mathbf{t} = \mathrm{col}(t_1, \ldots, t_m)$, $H(\mathbf{t}) = \sum_{i=1}^m h_i(t_i)$. Then problem (1) could be written in decentralized way as follows

$$\min_{\mathbf{t}} \sum_{i=1}^m h_i(t_i)$$

$$\text{s.t. } \mathbf{Wt} = 0.$$

Its dual form is

$$\max_{\mathbf{t}}\{\langle \mathbf{z}, \mathbf{W}\mathbf{t}\rangle - H(\mathbf{t})\} = -\min_{\mathbf{z}} H^*(\mathbf{W}^\top \mathbf{z}),$$

and the gradient of the objective can be computed using Demyanov'Danskin's theorem:

$$\nabla H^*(\mathbf{z}) = \arg\max_{\mathbf{t}}\{\langle \mathbf{z}, \mathbf{t}\rangle - F(\mathbf{E}\mathbf{t})\}.$$

From smaller dimension of t comparing to x we can expect that computation of $\nabla H^*(\mathbf{z})$ is easier than calling conventional first-order dual oracle, the only drawback is the necessity of storing matrix E and performing multiplications by E.

Let μ_t and L_t be the constants of strong convexity and Lipschitz smoothness of h_i respectively for all $i = 1, \ldots, m$. Then, obviously, $\mu_t \geq \mu$ and $L_t \leq L$. For example, if $f_i(x)$ is twice continuously differentiable, then its smoothness constant can be computed as $L_{x,i} = \sup_{x \in \mathbb{R}^n} \lambda_{\max}(\nabla^2 f_i(x))$. The smoothness constant of $h_i(t)$ is given by $L_{t,i} = \sup_{t \in \mathbb{R}^{d_t}} \lambda_{\max}(E^\top \nabla^2 f_i(Et)E)$. Note that the dimension of t can be computed as $d_t = d - \mathrm{rank}(B)$. In the latter variant the maximum is taken over a smaller set of points, and multiplication by E is likely to further reduce the smoothness constant (and increase strong convexity constant).

Since $H(\mathbf{t})$ is L_t-smooth and μ_t-strongly convex, we have that $F^*_{[\mathbf{Bx}=0]}(\mathbf{z}) = H^*(\mathbf{z})$ is $\frac{1}{\mu_t}$-smooth and $\frac{1}{L_t}$-strongly convex [6].

Thus, the fast gradient method [15] applied to the dual problem requires

$$O\left(\sqrt{\frac{L_t}{\mu_t}}\chi(W)\log\frac{1}{\varepsilon}\right),$$

dual-oracle calls and communication rounds to ensure $F(\mathbf{x}^N) - F(\mathbf{x}) \leq \varepsilon$ and $\|\mathbf{A}\mathbf{x}^N\| \leq \varepsilon$ (see Sect. 5 for details). And using Chebyshev acceleration as described in Sect. 4 we can reduce the complexities to

$$N = O\left(\sqrt{\frac{L_t}{\mu_t}}\log\frac{1}{\varepsilon}\right) \text{ oracle calls at each node,}$$

$$O\left(N\sqrt{\chi(W)}\right) \text{ communications.}$$

7   Numerical Experiments

In the simulation we consider the following smooth, strongly convex objective function:

$$f_i(x) = \frac{1}{2}\|C_i x - d_i\|_2^2 + \frac{\theta}{2}\|x\|_2^2,$$

$$F(\mathbf{x}) = \frac{1}{2}\|\mathbf{C}\mathbf{x} - \mathbf{d}\|_2^2 + \frac{\theta}{2}\|\mathbf{x}\|_2^2,$$

$$\mathbf{C} = \operatorname{diag}(C_1, \ldots, C_m), \ \mathbf{d} = \operatorname{col}(d_1, \ldots, d_m).$$

We consider different parameters of the problem such as the dimension of x, the rank of $B \in \mathbb{R}^{\dim(x) \times \dim(x)}$ and the number of nodes. For each case we plot function error and constraints violation norm at each iteration for all our algorithms: APDG, Locally and Globally Dual approaches. The Chebyshev acceleration is not applied in the experiments, so each iteration corresponds to one gradient computation (gradient of primal function in case of APDG, and gradient of dual function is case of dual approaches). We also provide tables with comparison of time and number of iterations required to achieve given accuracy. Time is measured with our Python/NumPy [5] implementation of the algorithms, which is available on GitHub[1].

1. For the first case we consider the ring network with $m = 5$ nodes, $x \in \mathbb{R}^{40}$ and rank $B = 1$. Typical convergence plot is shown on Fig. 1. One can see that all algorithms converge linearly, with the fastest one in terms of iterations number being Locally Dual, and the slowest one being APDG. However, computing the gradient of a dual function might be an arithmetically more expensive operation than computing primal gradient in the black-box scenario. In our implementation we compute the gradient of dual function by numerically solving the system of linear equations with its right-hand part being changed between iterations. It means that one iteration of the Dual methods is more time-consuming than one iteration of APDG. In the Table 1, we compare computational time and number of iterations required to achieve given accuracy. The results are averaged for 100 randomly generated problems.

Table 1. Time and iterations for achieving $\|\mathbf{A}\mathbf{x}^k\| < 10^{-2}$. Averaged over 100 experiments. Problem parameters: 5 nodes, $\dim(x) = 40$, rank $B = 1$.

	APDG	Globally dual	Locally dual
Iterations	875.3	502.7	276.7
Time (s)	0.193	0.510	0.233

---

[1] Source code: https://github.com/niquepolice/decentr_constr_dual.

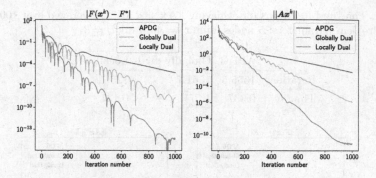

Fig. 1. 5 nodes, $\dim(x) = 40$, rank $B = 1$.

2. Next we use the same number of nodes and the dimension of x, but increase the rank of B. Even for rank $B = 3$ the condition number of the locally dual problem usually is about two orders of magnitude smaller than the condition number of the globally dual problem, therefore the globally dual approach has a significant advantage in that case. Typical convergence plots are shown in Fig. 2, averaged iteration and time complexities for satisfying stopping criteria are shown in Table 2.

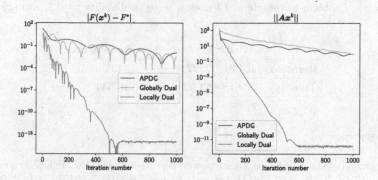

Fig. 2. 5 nodes, $\dim(x) = 40$, rank $B = 3$.

3. In the case of higher dimension (10 nodes, $\dim(x) = 100$, rank $B = 1$) we used Erdős-Rényi random communication graphs with edge probability $= 0.3$. APDG seems to converge much faster by constraints violation norm at first iterations then other methods (Fig. 3), and its convergence rate is close to other methods. See also Table 3 for averaged results of multiple experiments.

Table 2. Time and iterations for achieving $\|\mathbf{A}x^k\| < 10^{-1}$. Averaged over 100 experiments. Problem parameters: 5 nodes, $\dim(x) = 40$, rank $B = 3$.

	APDG	Globally dual	Locally dual
Iterations	1555.5	1551.7	123.1
Time (s)	0.337	1.577	0.127

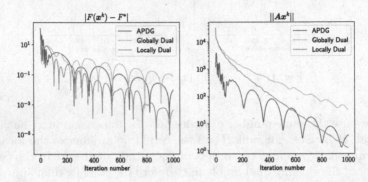

Fig. 3. Erdős-Rényi graph on 10 nodes, average degree $= 3.6$. $\dim(x) = 100$, rank $B = 1$.

Table 3. Time and iterations for achieving accuracy $\|\mathbf{A}x^k\| < 10^1$. Averaged over 10 experiments. Problem parameters: 10 nodes, edge probability $= 0.3$, $\dim(x) = 100$, rank $B = 1$.

	APDG	Globally dual	Locally dual
Iterations	404.3	2227.9	1425.5
Time (s)	2.561	54.024	16.544

References

1. Erseghe, T.: Distributed optimal power flow using ADMM. IEEE Trans. Power Syst. **29**(5), 2370–2380 (2014). https://doi.org/10.1109/TPWRS.2014.2306495
2. Falsone, A., Margellos, K., Garatti, S., Prandini, M.: Dual decomposition for multi-agent distributed optimization with coupling constraints. Automatica **84**, 149–158 (2017)
3. Falsone, A., Notarnicola, I., Notarstefano, G., Prandini, M.: Tracking-ADMM for distributed constraint-coupled optimization. Automatica **117**, 1–13 (2020)
4. Gorbunov, E., Dvinskikh, D., Gasnikov, A.: Optimal decentralized distributed algorithms for stochastic convex optimization. arXiv:1911.07363 (2019)
5. Harris, C.R., et al.: Array programming with NumPy. Nature **585**(7825), 357–362 (2020). https://doi.org/10.1038/s41586-020-2649-2
6. Kakade, S.M., Shalev-Shwartz, S., Tewari, A.: On the duality of strong convexity and strong smoothness: learning applications and matrix regularization. Technical report. Toyota Technological Institute (2009)
7. Kovalev, D., Gasnikov, A., Richtárik, P.: Accelerated primal-dual gradient method for smooth and convex-concave saddle-point problems with bilinear coupling. arXiv preprint arXiv:2112.15199 (2021)

8. Kovalev, D., Salim, A., Richtárik, P.: Optimal and practical algorithms for smooth and strongly convex decentralized optimization. Adv. Neural. Inf. Process. Syst. **33**, 18342–18352 (2020)

9. Liang, S., Wang, L.Y., Yin, G.: Distributed smooth convex optimization with coupled constraints. IEEE Trans. Autom. Control **65**, 347–353 (2020)

10. Liang, S., Zheng, X., Hong, Y.: Distributed ninsmooth optimization with coupled inequality constraints via modified Lagrangian function. IEEE Trans. Autom. Control **63**, 1753–1759 (2018)

11. Molzahn, D.K., et al.: A survey of distributed optimization and control algorithms for electric power systems. IEEE Trans. Smart Grid **8**(6), 2941–2962 (2017). https://doi.org/10.1109/TSG.2017.2720471

12. Necoara, I., Nedelcu, V.: Distributed dual gradient methods and error bound conditions. arXiv:1401.4398 (2014)

13. Necoara, I., Nedelcu, V.: On linear convergence of a distributed dual gradient algorithm for linearly constrained separable convex problems. Automatica **55**, 209–216 (2015). https://doi.org/10.1016/j.automatica.2015.02.038, https://www.sciencedirect.com/science/article/pii/S0005109815001004

14. Necoara, I., Nedelcu, V., Dumitrache, I.: Parallel and distributed optimization methods for estimation and control in networks. J. Process Control **21**(5), 756–766 (2011). https://doi.org/10.1016/j.jprocont.2010.12.010, https://www.sciencedirect.com/science/article/pii/S095915241000257X. special Issue on Hierarchical and Distributed Model Predictive Control

15. Nesterov, Y.: A method of solving a convex programming problem with convergence rate $o(1/k^2)$. Soviet Math. Doklady **27**(2), 372–376 (1983)

16. Nesterov, Y.: Introductory Lectures on Convex Optimization: A Basic Course. Kluwer Academic Publishers, Massachusetts (2004)

17. Patari, N., Venkataramanan, V., Srivastava, A., Molzahn, D.K., Li, N., Annaswamy, A.: Distributed optimization in distribution systems: use cases, limitations, and research needs. IEEE Trans. Power Syst. 1–1 (2021). https://doi.org/10.1109/TPWRS.2021.3132348

18. Rostampour, V., Haar, O.T., Keviczky, T.: Distributed stochastic reserve scheduling in AC power systems with uncertain generation. IEEE Trans. Power Syst. **34**(2), 1005–1020 (2019). https://doi.org/10.1109/TPWRS.2018.2878888

19. Scaman, K., Bach, F., Bubeck, S., Lee, Y.T., Massoulié, L.: Optimal algorithms for smooth and strongly convex distributed optimization in networks. In: Precup, D., Teh, Y.W. (eds.) Proceedings of the 34th International Conference on Machine Learning. Proceedings of Machine Learning Research, vol. 70, pp. 3027–3036. PMLR, International Convention Centre, Sydney, Australia (2017). http://proceedings.mlr.press/v70/scaman17a.html

20. Wang, J., Hu, G.: Distributed optimization with coupling constraints in multicluster networks based on dual proximal gradient method. arXiv preprint arXiv:2203.00956 (2022)

21. Wnag, Z., Ong, C.J.: Distributed model predictive control of linear descrete-times systems with local and global cosntraints. Automatica **81**, 184–195 (2017)

22. Yarmoshik, D., Rogozin, A., Khamisov, O., Dvurechensky, P., Gasnikov, A., et al.: Decentralized convex optimization under affine constraints for power systems control. arXiv preprint arXiv:2203.16686 (2022)

23. Yuan, D., Ho, D.W.C., Jiang, G.P.: An adaptive primal-dual subgradient algorithm for online distributed constrained optimization. IEEE Trans. Cybern. **48**, 3045–3055 (2018)

Game Theory and Mathematical Economics

Analysis of the Model of Optimal Expansion of a Firm

Anna Flerova[ID] and Aleksandra Zhukova[✉][ID]

Federal Research Center "Computer Science and Control"
of the Russian Academy of Sciences, Vavilov str, 44, bld. 2, Moscow, Russia
sasha.mymail@gmail.com
https://www.frccsc.ru

Abstract. In this paper we study a production expansion problem in different cases, including stochastic. The problem for the producer is to choose optimal way of investment in the production expansion. Different optimal control methods to study these cases are reviewed and the limits of their applicability are considered. We extend the analysis to the case of stochastic gain for investment, where the problem is set in continuous and discrete time with various assumptions. For the discrete time version we find the exact solution which has the same structure as in the deterministic model. In the continuous time we currently present the asymptotic analysis and study the specifics of the solution with respect to the terminal value.

Keywords: Production expansion · Optimal control · HJB equation

1 Introduction

This work is a study of the investment behavior of an enterprise (or a producer), influenced by various factors. The constructed mathematical model is formalized by different types of optimal control problems that require different solution techniques. The key factors that describe the environment might be deterministic or stochastic, and the model might be considered in continuous and discrete time. The influence of model parameters on the optimal solution is studied. Various versions of the model are considered and compared.

In this paper we consider the problem of optimal investment in production expansion [1–3]. The original baseline model [1,2] described the problem of profit maximization in the form we outline in Sect. 2. The authors demonstrate the existence of the piecewise-continuous control and optimality of the control using the Pontryagin's maximum principle.

This model admits several extensions. Firstly, the tax paid by the company might follow nonlinear relationship with the company's profits [2]. The author suggested to consider progressive tax so that the integrand in the goal functional is a concave function of the profit. This assumption leads to a more smooth optimal investment, compared to the baseline model where the investment jumps

N. Olenev et al. (Eds.): OPTIMA 2022, CCIS 1739, pp. 109–123, 2022.
https://doi.org/10.1007/978-3-031-22990-9_8

from maximum investment to no investment regime. Another extension is introduced in a more recent paper [4]. The authors add uncertainty assuming lack of information about the market conditions parameter α. The authors analyse the control in the form of an algorithm that delivers sub-optimal solution.

Several new modifications of the model are introduced in this paper. Firstly, we assume the possibility of uncertainty in the form of stochastic gain for investment and reformulate the model in discrete time to be able to analyse it in the simplest case. We further extended the model by adding Brownian motion component to the dynamics of the income of the producer. We demonstrate that the solution to the modified model is qualitatively similar to the deterministic version if we impose the information constraint that the investment decision precedes the realization of the gain in income resulting from the investment.

This study is important from both theoretical point of view and economic analysis. The former is valuable since we present the approach to the description of the stochastic optimal control and its analysis. As for the economic sense of the model, it is relevant as the authors of [5] remark, in the current conditions, investment is typically made out of the income of owners and it is important to analyze the risks associated with this process. From the economic policy point of view, it is important to understand the consequences of increased uncertainty.

This work contributes to modelling the producer's decision-making in the conditions of varying revenue and other external conditions [6–8]. This model might also include the aspect of uncertainty in the line of studies of the more realistic description of the production where the owners work on an imperfect capital markets [9] with different rates for deposits and loans.

2   Formulation of the Problem

Let us analyze a problem formulated in [1]. A firm produces a product, receives $x(t)$ of money per unit of time for the sale of this product. The share $u(t)$ of the output $x(t)$ is used to expand production:

$$\frac{dx}{dt} = \alpha u(t)x(t),$$

where α is the parameter that characterizes the return on investment. It might be affected by the economic environment, the demand for the firm's product or some other external or internal circumstances.

We assume that there are no transaction costs, sales occur instantly.

The initial capital is known: $x(0) = x_0 > 0$. We assume also that the production costs are proportional to the output produced with coefficient β (per unit of time the production costs are equal to $\beta x(t)$). Suppose that the producer pays taxes in proportion to earnings. The portion b_1 of the resulting profit is paid as a tax.

Therefore, the producer's profit from time $t = 0$ to $t = T$ (where T is known) is equal to

$$Q = \int_0^T (1 - b_1)(x(t) - u(t)x(t) - \beta x(t))dt.$$

Let us denote $\mu = 1 - \beta$, It is clear that under the assumption that the producer pays taxes in proportion to earnings the parameter $1 - b_1$ has no effect on the maximum of the functional (1). So, further in this work we will consider $b_1 = 0$. So, the problem is to choose the control $u(t)$ that gives the maximum profit:

$$\int_0^T (\mu - u(t))x(t)dt \to max \qquad (1)$$

$$\dot{x}(t) = \alpha u(t)x(t), \qquad (2)$$

$$0 \leqslant u(t) \leqslant \mu, \qquad (3)$$

$$x(0) = x_0. \qquad (4)$$

3   The Deterministic Model of Optimal Expansion of the Firm

Let us begin with solving the producer's problem (1)–(4) in continuous time under the condition that the ratio α (the parameter that characterizes the return on investment) is a positive constant $\alpha = const > 0$. We will refer to this analysis as we move on to the stochastic version. This way this problem is a classical optimal control problem and one can solve it using the Pontryagin's maximum principle, and in this study the Hamilton-Jacobi-Bellman function will be used to handle with the problem.

Theorem 1. *The optimal control function for the problem (1)–(4), where α is a positive constant, is*

$$if \quad \tau > 0, \quad u^{opt}(t) = \begin{cases} \mu, & if \quad 0 \leqslant t \leqslant \tau, \\ 0, & if \quad \tau < t \leqslant T; \end{cases}$$

$$if \quad \tau \leqslant 0, \quad u^{opt}(t) = 0, \quad t \in [0,T],$$

where $\tau = T - \frac{1}{\alpha\mu}$.

Proof. The necessary conditions in the form of the Hamilton-Jacobi-Bellman equation for the value function $S(t,x)$ of the problem (1)–(4) are

$$-\frac{\partial S(t,x)}{\partial t} = \max_{0 \leqslant u \leqslant \mu} \left[\mu x(t) - u(t)x(t) + \alpha u(t)x(t)\frac{\partial S(t,x)}{\partial x} \right],$$

with the boundary condition $S(T,x) = 0$, that is

$$-\frac{\partial S}{\partial t} = \mu x + \max_{0 \leqslant u \leqslant \mu} \left[\left(\alpha\frac{\partial S}{\partial x} - 1 \right) xu \right]. \qquad (5)$$

So, the value of the function $u(t)$ that maximizes the expression in the square brackets in (5) is

$$u(t,x) = \begin{cases} 0, & if \quad \alpha\frac{\partial S}{\partial x} < 1, \quad (II) \\ \mu, & if \quad \alpha\frac{\partial S}{\partial x} > 1. \quad (I) \end{cases} \qquad (6)$$

Let us study these two cases. In case I when $u(t) = \mu$ Eq. (5) becomes

$$-\frac{\partial S}{\partial t} = \alpha \frac{\partial S}{\partial x} \mu x. \tag{7}$$

This equation is a linear homogeneous first-order partial differential equation. The solution of the Eq. (7) may be presented as $S(t, x) = F(xe^{-\mu \alpha t})$, where F is a continuous function. If case I executes at the end of the considered interval (at some interval $[t_1, T]$, where $0 < t_1 < T$), then the boundary condition $S(T, x) = F(xe^{-\mu \alpha T}) = 0$ must be satisfied for any x. But it is possible only if $F \equiv 0$, which is a trivial solution, when the functional remains zero.

In case II when $u(t) = 0$ Eq. (5) becomes $-\partial S/\partial t = \mu x$.

In this case the value function is decreasing and can reach the boundary condition $S(T, x) = 0$. So, at some interval $[\tau, T]$, where $0 < \tau < T$, $S(t, x) = \mu x(T - t)$. The switching time τ may be found from (6), which says that the optimal control switches when $\frac{\partial S}{\partial x} = \frac{1}{\alpha}$:

$$\alpha \left. \frac{\partial S(t, x)}{\partial x} \right|_{t=\tau} = \alpha \mu (T - \tau) = 1$$

and

$$\tau = T - \frac{1}{\alpha \mu}. \tag{8}$$

Notice, that $\left. \frac{\partial^2 S(t,x)}{\partial x \partial t} \right|_{\tau} < 0$ and case II on the right from τ changes on case I on the left from τ.

The value function $S(t, x)$ has to be continuous at the point τ, so the following qualities hold:

$$F(xe^{-\mu \alpha \tau}) = \mu x(T - \tau) \quad \Rightarrow \quad F(xe^{-\mu \alpha T + 1}) = \frac{x}{\alpha}.$$

Let's substitute $xe^{-\mu \alpha T + 1} = y$, then $x = ye^{\mu \alpha T - 1}$ and $F(y) = \frac{ye^{\mu \alpha T - 1}}{\alpha}$. That is $S(t, x) \Big|_{t < \tau} = \frac{xe^{-\alpha \mu t}}{\alpha} e^{\alpha \mu T - 1} = \frac{x}{\alpha} e^{\mu \alpha (\tau - t)}$, which is also a decreasing function. So, the optimal control function has got not more than one point of change in regime.

It is easy to see that the obtained value function $S(t, x)$ is continuous and smooth.

The following important fact should be noted. The Hamilton-Jacobi-Bellman equation (5) for the problem (1)–(4) gives us not only necessary but also sufficient conditions of optimality [10].

So, the optimal control function for the problem (1)–(4) is

$$u^{opt}(t) = \begin{cases} \mu, & \text{if} \quad 0 \leqslant t \leqslant T - \frac{1}{\alpha \mu}, \\ 0, & \text{if} \quad T - \frac{1}{\alpha \mu} < t \leqslant T. \end{cases} \tag{9}$$

And the theorem is proved.

This theorem shows that the producer has two stages of lifetime. On the initial stage the producer spends all the income for the production extension and then on the second stage consumes all the income.

Notice that depending on the parameters (e.g., if the time T is not enough) there might not be the period of no investment when $u^{opt}(t) = \mu$ and producer is only consuming the income all the time. It is the case when $\tau = T - \frac{1}{\alpha\mu} < 0$. In general, the relation between this two stages depends on the parameters of the model T, α and μ.

If one introduces the discounting coefficient δ into our model, it turns into

$$\int_0^T (\mu - u(t)) x(t) e^{-\delta t} dt \tag{10}$$

$$\dot{x}(t) = \alpha u(t) x(t), \ x(0) = x_0, \ u \in [0, \mu]. \tag{11}$$

Using the Pontriagin's maximum principle one can show that the optimal control for this problem is

$$u_\delta(t) = \begin{cases} \mu, & t < \ln\left(\frac{\alpha\mu-\delta}{\alpha\mu}\right)\delta^{-1} + T, \\ 0 & t > \ln\left(\frac{\alpha\mu-\delta}{\alpha\mu}\right)\delta^{-1} + T. \end{cases} \tag{12}$$

The optimal control has not more then one switching point, and switches from μ to 0 at timepoint

$$\tau_\delta = \ln\left(\frac{\alpha\mu-\delta}{\alpha\mu}\right)\delta^{-1} + T, \tag{13}$$

if $\tau_\delta > 0$. For small δ we have $\tau_\delta \approx T - \frac{1}{\alpha\mu} - \frac{\delta}{2\alpha^2\mu^2} - \frac{\delta^2}{3\alpha^3\mu^3}$, which is the same as the model without discounting. *The role of the discounting parameter δ is to shorten the time of investment* (where investment rate is at maximum) and expand the period of spending (where there is no investment $u_\delta = 0$).

But let us note, that the introduction of the discounting parameter did not have a fundamentally affect on the form of the solution: the optimal control, as before, has no more than one switching point.

4 The Stochastic Model of Optimal Expansion of the Firm

4.1 Stochastic Model of Production Expansion in Discrete Time

Let's move to discrete time in order to take stochastics into account more easily and rewrite the producer's problem (1)–(4):

$$\sum_{t=0}^{T-1} (\mu - u_t) x_t \to \max \tag{14}$$

$$x_{t+1} = x_t + \alpha_t u_t x_t, \tag{15}$$

$$x(0) = x_0 > 0, \tag{16}$$

$$0 \le u_t \le \mu. \tag{17}$$

To account the uncertainty in investment let us assume ratio α to be a Bernulli random variable:

$$\alpha_t = \begin{cases} \alpha_1 \text{ with probability } p_1, \\ \alpha_2 \text{ with probability } p_2, \\ \quad ... \\ \alpha_n \text{ with probability } p_n, \end{cases} \qquad (18)$$

where $\sum_{i=1}^{n} p_i = 1$ and $\alpha_i > 0$ for $i = 1, ..., n$.

For the problem in the discrete time it appears to be very important to understand in what order the events take place. This was discussed in the paper on modelling the agents when a change of state occurs according to the Poisson process [11,12]. In that setting, it was essential whether the agent's state is described by the left-continuous or right continuous process. The resulting solution would depend on this assumption. A similar problem arises here.

The Bellman equation defines the necessary conditions for optimality of the investment expenditure u.

When u_t is defined before the realization of the α_t, the Bellman equation contains the maximum over the average expected value of the future state

$$S_t(x_t) = \max_{0 \leq u_t \leq \mu} \left[f(u_t, x_t) + \mathbb{E}_{\alpha_t} S_{t+1}(x_{t+1}) \right].$$

On the contrary, if the control u_t is defined *after* the agent knows the realized value of α_t, the Bellman equation would be different, having the expected maximum over the optimal choice of the u_t contingent on the α_t and the value of the future state

$$S_t(x_t) = \mathbb{E}_{\alpha_t} \max_{0 \leq u_t \leq \mu} \left[f(u_{t,\alpha_t}, x_t) + S_{t+1}(x_{t+1,\alpha_t}) \right]. \qquad (19)$$

One may see that the solution to the Eq. (19) defines the control as a complex function u_{t,α_t} contingent on every realization of the random process α_t.

We decided to start with the analysis of the first option, where the agent has to make the investment decision in the beginning of the time interval $[t, t + 1]$. It makes sense to assume the limited information about the realized gain characterized by the random α_t by the end of this period.

In the next section we present the analysis of the optimal control problem under this assumption.

4.2 Solution to Stochastic Model of Production Expansion in Discrete Time

Proposition 1. *There exists a solution to the HJB equation (4.1) in the form:*

$$u_t^{opt} = \begin{cases} \mu, & \text{if } t < T - 1 - \frac{1}{E(\alpha)\mu}, \\ 0, & \text{if } t > T - 1 - \frac{1}{E(\alpha)\mu}. \end{cases}$$

Proof. To simplify the expressions further in this subsection we assume that the ratio α_t (18) may take two values:

$$\alpha_t = \begin{cases} \alpha_1 \text{ with probability } & p, \\ \alpha_2 \text{ with probability } & 1 - p, \end{cases}$$

It is easy to verify that all following calculations are also valid for any number of values that α_t may take. The necessary conditions in the form of the Hamilton-Jacobi-Bellman equation for the value function $S_t(x)$ of the problem (14)–(17) are

$$S_t(x_t) = \max_{0 \leq u_t \leq \mu} [(\mu - u_t) x_t + \mathbb{E} S_{t+1}(x_{t+1})]$$

$$= \max_{0 \leq u_t \leq \mu} [(\mu - u_t) x_t + p S_{t+1}(x_t + \alpha_1 u_t x_t) + (1 - p) S_{t+1}(x_t + \alpha_2 u_t x_t)]$$

with the boundary condition $S_T(x) = 0$.

This problem is considered in the reverse time from the last moment in time to the first. At the end we have $S_T(x) = 0$, so at the previous step we assume

$$S_{T-1}(x_{T-1}) = \max_{0 \leq u_{T-1} \leq \mu} [(\mu - u_{T-1}) x_{T-1}].$$

Thus, $u_{T-1} = 0$ and $S_{T-1}(x_{T-1}) = \mu x_{T-1} = \mu x_T$.

Moving to the step $T - 2$:

$$S_{T-2}(x_{T-2}) = \max_{0 \leq u_{T-2} \leq \mu} [(\mu - u_{T-2}) x_{T-2} + p S_{T-1}(x_{T-2} + \alpha_1 u_{T-2} x_{T-2})$$

$$+ (1 - p) S_{T-1}(x_{T-2} + \alpha_2 u_{T-2} x_{T-2})]$$

$$= \max_{0 \leq u_{T-2} \leq \mu} [(\mu - u_{T-2}) x_{T-2} + (p + 1 - p) \mu x_{T-2} + \mathbb{E}(\alpha) \mu u_{T-2} x_{T-2}]$$

$$= \max_{0 \leq u_{T-2} \leq \mu} [x_{T-2}(2\mu + (\mathbb{E}(\alpha)\mu - 1) u_{T-2})].$$

Optimal control on the step $T - 2$ is

$$u_{T-2} = \begin{cases} \mu, \text{ if } \mathbb{E}(\alpha)\mu - 1 > 0, \\ 0, \text{ if } \mathbb{E}(\alpha)\mu - 1 < 0. \end{cases}$$

Suppose that $u_{T-2} = 0$ then we have $S_{T-2} = 2\mu x_{T-2} = 2\mu x_T$ and on the previous step $T - 3$ we have

$$S_{T-3}(x_{T-3}) = \max_{0 \leq u_{T-3} \leq \mu} [(\mu - u_{T-3}) x_{T-3} + p S_{T-2}(x_{T-3} + \alpha_1 u_{T-3} x_{T-3})$$

$$+ (1 - p) S_{T-2}(x_{T-3} + \alpha_2 u_{T-3} x_{T-3})]$$

$$= \max_{0 \leq u_{T-3} \leq \mu} [(\mu - u_{T-2}) x_{T-3} + (p + 1 - p) \mu x_{T-3} + \mathbb{E}(\alpha) \mu u_{T-2} x_{T-3}]$$

$$= \max_{0 \leq u_{T-3} \leq \mu} [x_{T-3}(3\mu + (2\mathbb{E}(\alpha)\mu - 1) u_{T-3})].$$

Let's suppose that there are k steps from the end for which $u_t = 0$, $t \geq T - k + 1$, then on the step $T - k$ we obtain the following value function:

$$S_{T-k}(x_{T-k}) = \max_{0 \leq u_{T-k} \leq \mu} \left[x_{T-k} \left(k\mu + ((k-1) \, \mathbb{E}(\alpha) \, \mu - 1) \, u_{T-k} \right) \right].$$

We have shown that $u_{T-1} = 0$. Let l ($l \in \mathbb{N}$, $l \geqslant 3$) be the smallest number that satisfies the inequality $(l-2) \, \mathbb{E}(\alpha) \, \mu - 1 > 0$. Notice that if $l > T$ then $u_t = 0$ every step of the life of the firm (t = 0, ..., T). But we will consider a more interesting case when $l < T$, then $u_t = 0$ for $t > T - l + 1$ and $u_{T-l+1} = \mu$. Such $T - l + 1$ is the first timepoint in the reverse time when the coefficient to u_{T-l+1} in S_{T-l+1} is positive and the optimal control u_{T-l+1} is nonzero. Following this

$$S_{T-l+1}(x_{T-l+1}) = x_{T-l+1}\mu \, (l-2) \, (1 + \mathbb{E}(\alpha) \, \mu).$$

Let's see the next (in the reverse time) step.

$$S_{T-l}(x_{T-l}) = \max_{0 \leqslant u_{T-l} \leqslant \mu} [x_{T-l}(\mu + \mu(l-2)(1 + \mathbb{E}(\alpha)\mu$$
$$+ \; (\mathbb{E}(\alpha)\mu(l-2)(1 + \mathbb{E}(\alpha)\mu) - 1)u_{T-l}].$$

Notice that

$$\mathbb{E}(\alpha) \, \mu \, (l-2) \, (1 + \mathbb{E}(\alpha) \, \mu) - 1 \; > \; (l-2) \, \mathbb{E}(\alpha) \, \mu - 1 \; > 0.$$

It means that $u_{T-l} = \mu$ and

$$S_{T-l}(x_{T-l}) = x_{T-l}\mu \, (l-2) \, (1 + \mathbb{E}(\alpha) \, \mu)^2$$

It is easy to check that for each next step to the beginning $u_{T-t} = \mu$ and

$$S_{T-t}(x_{T-t}) = \mu x_{T-t} \, (l-2) \, (1 + \mathbb{E}(\alpha) \, \mu)^{2+t}$$

for t from $l - 1$ to T.

Now we have got the optimal control for our problem:

$$u_t^{\text{opt}} = \begin{cases} \mu, & \text{if } (T - t - 1) \, E(\alpha) \, \mu - 1 \; > 0, \\ 0, & \text{if } (T - t - 1) \, E(\alpha) \, \mu - 1 \; < 0. \end{cases} \qquad (20)$$

The proposition is proved.

Important to note that if α is a positive constant and the problem (14)–(17) is deterministic, the optimal control (20) becomes

$$u_{\text{determ}}^{\text{opt}} = \begin{cases} \mu, & \text{if } \; 0 \leqslant t \leqslant T - 1 - \frac{1}{\alpha\mu}; \\ 0, & \text{if } \; T - 1 - \frac{1}{\alpha\mu} \leqslant t \leqslant T - 1. \end{cases}$$

And this optimal control looks like the optimal control (9), which brings the optimal solution to the continuous problem (1)–(4).

5   Stochastic Model of Production Expansion in Continuous Time

The stochastic version of this model in continuous time might assume the random process of income in the form of

$$dx_t = \alpha u_t x_t dt + \sigma x_t dB_t, \tag{21}$$

where B_t is a Brownian motion. In this formulation, $u_t = u(t, \omega)$ is a stochastic process. In order to be non-anticipating, that is, based on information up to time t, the function $\omega \to u(\omega, t)$ is assumed to be measurable with respect to sigma algebra $\mathcal{F}_t$ of natural filtration $\{\mathcal{F}_t\}$ and α is a positive constant again.

A possible variation is to assume random Poisson jumps of the income

$$dx_t = \alpha u_t x_t dt + \sigma x_t dN_t, \tag{22}$$

where N_t is the Poisson counting process. Our experience [11,12] shows that the analysis requires a different technique with Lagrange's multipliers methods. We are planning this variation for further research.

The agent's goal is augmented by a terminal term $F(x(T))$, whose role we demonstrate below

$$\mathbb{E}\left[\int_0^T (\mu - u(t))x(t)dt + F(x(T))\right] \to max \tag{23}$$

$$dx(t) = \alpha u(t)x(t)dt + \sigma x(t)dB_t, \tag{24}$$

$$u \in [0, \mu], \; x(0) = x_0. \tag{25}$$

5.1   The Production Expansion Problem with a Linear Terminal Component

Let's briefly consider a deterministic version of the problem with the terminal component:

$$\int_0^T (\mu - u(t))x(t)dt + kx(T) \to max \tag{26}$$

$$\dot{x}(t) = \alpha u(t)x(t), \tag{27}$$

$$0 \leqslant u(t) \leqslant \mu, \; x(0) = x_0. \tag{28}$$

Proposition 2. *The optimal control function for the problem (26)–(28) is*

$$if \quad k < \frac{1}{\alpha}, \quad u^{opt}(t) = \left\{ \begin{array}{ll} \mu, & if \;\; 0 \leqslant t \leqslant T - \frac{1}{\alpha\mu} + \frac{k}{\mu}, \\ 0, & if \;\; T - \frac{1}{\alpha\mu} + \frac{k}{\mu} < t \leqslant T, \end{array} \right.$$

$$otherwise \quad u^{opt}(t) = \mu \quad for \quad t \in [0, T].$$

Proof. The Pontryagin maximum principle states the optimal control to maximize the Hamilton-Pontryagin function $H(t, x, u, p) = (\mu - u)x + p\alpha u x$, where the conjugate variable $p(t)$ satisfies the equation $\dot{p}(t) = u - \mu - p\alpha u$ and the transversality condition $p(T) = k$. From that the optimal control is

$$u(t) = \begin{cases} \mu, & \text{if} \quad p(t) > \frac{1}{\alpha}, \\ 0, & \text{if} \quad p(t) < \frac{1}{\alpha}. \end{cases} \tag{29}$$

If $u(t) = 0$ then $\dot{p}(t) = -\mu$ and $p(t)$ is a decreasing function. If $u(t) = \mu$ then $\dot{p}(t) = -p\alpha\mu$ and $p(t)$ is also a decreasing function. So the optimal control has got not more than one switching point and it depends on the value of k. If $k > \frac{1}{\alpha}$ in respect that $p(T) = k$ there is no switch point and optimal control is $u(t) = \mu$ for $t \in [0, T]$. This means that the producer invests everything is possible to make the earnings $x(T)$ at the last time T the largest. If $k < \frac{1}{\alpha}$ the switching moment is easily found: $\tau = T - \frac{1}{\alpha\mu} + \frac{k}{\mu}$. So the optimal control for the problem (26)–(28) is:

$$\text{if} \quad k < \frac{1}{\alpha}, \quad u^{opt}(t) = \begin{cases} \mu, & \text{if} \quad 0 \leqslant t \leqslant T - \frac{1}{\alpha\mu} + \frac{k}{\mu}, \\ 0, & \text{if} \quad T - \frac{1}{\alpha\mu} + \frac{k}{\mu} < t \leqslant T, \end{cases} \tag{30}$$

otherwise $u^{opt}(t) = \mu$ for $t \in [0, T]$.

It is clear that if k is big enough the terminal component in (26) may change the form of the solution: the producer might have an investing period for all the interval $[0, T]$. The conditions of this problem allow him to get the maximum profit due to the last moment of time T. But if k is small enough, the solution (30) is similar to the solution of the problem (1)–(4).

6   Stochastic Model of Production Expansion in Continuous Time

Now returning to the stochastic model (32)–(34). The necessary conditions for optimal control u might be formulated for the value function

$$S(s, y) = \sup_{u} J^u(s, y), \tag{31}$$

where

$$J^u(s, y) = \mathbb{E}\left[\int_s^T (\mu - u(t))x(t)dt + \phi(x(T))\right] \tag{32}$$

$$dx(t) = \alpha u(t)x(t)dt + \sigma x(t)dB_t, \tag{33}$$

$$u \in [0, \mu], \quad x(s) = y. \tag{34}$$

The Hamilton-Jacobi-Bellman equation for this problem [13]

$$\sup_{v \in [0, \mu]} \left\{ \frac{\partial}{\partial s}S(s, y) + (\mu - v)y + \alpha v y \frac{\partial}{\partial y}S(s, y) + \frac{\sigma^2 y^2}{2}\frac{\partial^2}{\partial y^2}S(s, y) \right\} = 0. \tag{35}$$

The terminal condition

$$S(T, y) = \phi(y). \tag{36}$$

The optimal control value in (35) is similar to the (6)

$$v(t,x) = \begin{cases} 0, & \text{if } \alpha \frac{\partial S}{\partial x} < 1, & (A) \\ \mu, & \text{if } \alpha \frac{\partial S}{\partial x} > 1, & (B) \\ [0, \mu], & \text{if } \alpha \frac{\partial S}{\partial x} = 1. & (C) \end{cases} \tag{37}$$

6.1   Case (A)

For the case (A) the Eq. (35) turns into

$$-\frac{\partial}{\partial s} S(s, y) = \mu y + \frac{\sigma^2 y^2}{2} \frac{\partial^2}{\partial y^2} S(s, y). \tag{38}$$

One may try to solve this equation in the form

$$S(s, y) = H(s, y) - \frac{\mu y \ln(y)}{\sigma^2} + C_1 y + C_2, \tag{39}$$

which leads to the equation

$$-\frac{\partial}{\partial s} H(s, y) = \frac{y^2 \sigma^2}{2} \frac{\partial^2}{\partial y^2} H(s, y). \tag{40}$$

This equation might be solved in the form $H(s, y) = F_1(s) F_2(y)$. As a result the nonzero solution of (40) reduces to solving a system of equations

$$\frac{\sigma^2 y^2}{2 F_2(y)} \frac{d^2}{dy^2} F_2(y) = c = -\frac{1}{F_1(s)} \frac{d}{ds} F_1(s). \tag{41}$$

The case of $c = 0$. In the case of $c = 0$ the solution to the system (41) is

$$F_1(s) = const_1, \quad F_2(y) = y \, const_2 + const_3.$$

Returning to the (39) the solution $S(s, y)$ is still

$$S(s, y) = -\frac{\mu y \ln(y)}{\sigma^2} + \tilde{C}_1 y + \tilde{C}_2. \tag{42}$$

The Case of $c \neq 0$. This option leads to the solutions of the (41):

$$F_1(s) = C_3 e^{-cs}, \tag{43}$$

$$F_2(y) = C_4 y^{\frac{1}{2} + \frac{\sqrt{\sigma^2 + 8c}}{2\sigma}} + C_5 y^{\frac{1}{2} - \frac{\sqrt{\sigma^2 + 8c}}{2\sigma}}. \tag{44}$$

The (39) has the form of $S(s, y)$ as

$$S(s, y) = \tilde{C}_3 e^{-cs} \left(y^{\frac{1}{2} + \frac{\sqrt{\sigma^2 + 8c}}{2\sigma}} + \tilde{C}_4 y^{\frac{1}{2} - \frac{\sqrt{\sigma^2 + 8c}}{2\sigma}} \right) - \frac{\mu y \ln(y)}{\sigma^2} + \tilde{C}_1 y + \tilde{C}_2. \tag{45}$$

6.2   Case (B)

The case (B) leads to the equation (35) in the form

$$-\frac{\partial}{\partial s}S(s,y) = \alpha\mu y\frac{\partial}{\partial y}S(s,y) + \frac{\sigma^2 y^2}{2}\frac{\partial^2}{\partial y^2}S(s,y).\tag{46}$$

One may also solve it in the form $S(s,y) = F_1(s)F_2(y)$. Substituting into (46)

$$\frac{\sigma^2 y^2}{2F_2(y)}\frac{d^2}{dy^2}F_2(y) + \frac{y\mu\alpha}{F_2(y)}\frac{d}{dy}F_2(y) = c = -\frac{1}{F_1(s)}\frac{d}{ds}F_1(s).\tag{47}$$

The solution delivers

$$S(s,y) = C_1 e^{-cs} y^{1/2+1/2\frac{-2\alpha\mu+\sqrt{4\alpha^2\mu^2-4\alpha\mu\sigma^2+\sigma^4+8c\sigma^2}}{\sigma^2}}$$
$$+ C_2 e^{-cs} y^{1/2-1/2\frac{2\alpha\mu+\sqrt{4\alpha^2\mu^2-4\alpha\mu\sigma^2+\sigma^4+8c\sigma^2}}{\sigma^2}}.\tag{48}$$

We present the two solutions to (35) and (46), one of which should satisfy the terminal condition (36). If the function $\phi(y)$ were zero, as in the initial formulation of the problem (1)–(4), there would be no nontrivial solution of this form. Further analysis is needed to find the solution in the form different from the one attempted above.

If one assumes some nonzero $\phi(y)$ that enables to satisfy the terminal condition (36), another challenge appears to match the two solutions in the points where (37) switches from one value to the other

$$\alpha\frac{\partial}{\partial y}S(s,y) = 1.\tag{49}$$

The (42) contains the $y\ln(y)$ term with no undefined constants, whereas the (48) has only power functions of y with constants. Therefore, the (49) would be a complex equation that we plan to analyze in the future.

6.3   Approximate Asymptotic Solution to the Producer's Problem

Now we are able to present the asymptotic approximation for $S(s,y)$. For this, we substitute

$$y^2\frac{\partial^2}{\partial y^2}S(s,y) = S_1(s,y).\tag{50}$$

We start form the Eq. (35) where we have

$$-\frac{\partial}{\partial s}S(s,y) = \mu y + \frac{\sigma^2}{2}S_1(s,y).\tag{51}$$

This allows us to solve this equation for $S(s,y)$ to obtain at time interval $(s,T]$

$$S(s,y) = S(T,y) + \mu y(T-s) + \frac{\sigma^2}{2}\int_s^T y^2\frac{\partial^2}{\partial y^2}S(\tau,y)\,d\tau.\tag{52}$$

By substituting this expression into the right hand side of itself, we obtain

$$S(s,y) = S(T,y) + \mu y (T-s) + 1/2\,\sigma^2 y^2 \left(\frac{\partial^2}{\partial y^2} S(T,y)\right)(T-s)$$

$$+ 1/2\,\sigma^4 y^2 \int_s^T \int_\tau^T \frac{\partial^2}{\partial y^2} S(h,y)\,dh\,d\tau$$

$$+ \sigma^4 y^3 \int_s^T \int_\tau^T \frac{\partial^3}{\partial y^3} S(h,y)\,dh\,d\tau + \frac{\sigma^4 y^4}{4} \int_s^T \int_\tau^T \frac{\partial^4}{\partial y^4} S(h,y)\,dh\,d\tau. \quad (53)$$

Assuming the parameter σ to be small and the coefficient of the σ^4 to be bounded (the assumption to be verified), we obtain the order of σ^2 approximation

$$S(s,y) = S(T,y) + \mu y (T-s) + \frac{\sigma^2 y^2}{2} \frac{\partial^2}{\partial y^2} S(T,y)(T-s). \quad (54)$$

Using the same idea, one may rewrite the (46) equation in the form for some moment of time τ

$$S(s,y) = 1/2\,\sigma^2 y^2 e^{-2\alpha\mu s} \int_s^\tau e^{2h\alpha\mu} D_{2,2}(S)\left(h, y e^{-\alpha\mu s + h\alpha\mu}\right) dh + F_1\left(e^{-\alpha\mu s} y\right). \quad (55)$$

Here by $D_{2,2}(S)$ we denote the derivative with respect to the second argument.

By substituting it in the right-hand side of itself and leaving only the terms up to the order σ^2 and less, we come to the following form of the asymptotic expression

$$S(s,y) = F_1\left(e^{-\alpha\mu s} y\right) + 1/2\,\sigma^2 y^2 e^{-2\alpha\mu s} F_1''\left(e^{-\alpha\mu s} y\right)(\tau - s). \quad (56)$$

As in the deterministic case described in the beginning of this paper, we may also find the unknown function F_1 from the continuity of $S(s,y)$ at some point $s = \tau$. Therefore

$$F_1(z) = S(T, z e^{\alpha\mu\tau}) + \mu z e^{\alpha\mu\tau} T - \mu z e^{\alpha\mu\tau} \tau$$

$$+ \frac{\sigma^2 z^2}{2} (D_{2,2})(S)(T, z e^{\alpha\mu\tau})(T - \tau) e^{2\alpha\mu\tau}. \quad (57)$$

We demonstrate the approach on the particular case in the next section.

6.4   The Asymptotic Solution to the Producer's Problem with $\phi(x) = kx$

To illustrate the solution obtained in the previous section and to compare it to the deterministic model (26)–(28) we consider the case of $\phi(x) = kx$. Therefore,

$$S(T,y) = ky. \quad (58)$$

From (54) we have the value function for different time intervals

$$S(s,y) = ky + \mu yT - \mu ys, \tag{59}$$

$$S(s,y) = \mu y (T - \tau) e^{-\alpha \mu s + \alpha \mu \tau} + kye^{-\alpha \mu s + \alpha \mu \tau}, \tag{60}$$

that are derived based on assumption of taking the same value at time $t = \tau$. One may further see that

$$\frac{\partial}{\partial y} S(s,y) = k + \mu T - \mu s, \tag{61}$$

$$\frac{\partial}{\partial y} S(s,y) = \mu (T - \tau) e^{-\alpha \mu s + \alpha \mu \tau} + ke^{-\alpha \mu s + \alpha \mu \tau}. \tag{62}$$

Assuming that

$$\frac{\partial}{\partial y} S(\tau,y) = \frac{1}{\alpha}, \tag{63}$$

we obtain the value τ of the moment of time when these two functions become equal to each other.

$$\frac{\partial}{\partial y} S(\tau,y) = k + \mu T - \mu \tau, \quad \frac{\partial}{\partial y} S(\tau,y) = \mu (T - \tau) + k. \tag{64}$$

The value of τ corresponds to the deterministic version (26)–(28)

$$\tau = \left(k - \alpha^{-1} + \mu T \right) \mu^{-1}. \tag{65}$$

It is clear that this value is in the range $[0, T]$ only in the case of $k < 1/\alpha$.

One may verify that both (64) are decreasing functions of time that take the value of $1/\alpha$ only once. This defines the control (37) of the form identical to the optimal control (30) in the deterministic version of the model.

7   Conclusion

This paper presents the analysis of the model of optimal production expansion. The model is modified to take into account uncertainty in the environment that affects the future income for the producer. The influence of model parameters on the optimal solution is studied. Various versions of the model are considered and compared. Two possible stochastic versions are presented and analyzed: a discrete-time version and the continuous-time version with income having the Browninan motion component. In order to study the continuous-time problem, a generalization of the baseline model is introduced, that adds a terminal term to the producer's goal. We present an asymptotic approximation of the stochastic model, that demonstrates correspondence to the solution in the deterministic case. This work demonstrates that the solution to the model with uncertainty is qualitatively similar to the deterministic version and relies on the expected value of the uncertain parameter. The small uncertainty in the continuous time model requires adding a special terminant term to the goal functional and in the case of linear terminant the approximate solution is still similar to the deterministic model.

Acknowledgment. The research by Aleksandra Zhukova in section 6 was supported by RSCF grant No. 22-21-00746.

References

1. Arutyunov, A.V., Magaril-Il'yayev, G.G., Tihomirov, V.M.: Pontryagin's Maximum Principle. The Proof and Applications, Faktorial Press, Moscow (2006). (Russian)
2. Matveev, A.S., Yakubovich, V.A.: Optimal systems of control: ordinary differential equations. Special problems. SPb.: Punlishing House of the Saint-Petersburg University (2003). (Russian)
3. Pervozvanski, A.A.: Mathematical Models in Control of Production. Nauka, Moscow (1975)
4. Vedyakov, A.A., Vorobev, V.S., Tertychny-Dauri, V.Y.: Adaptive problem of extended reproduction with minimization of generalized costs. Sci. Tech. J. Inf. Technol. Mech. Opt. **130**(6), 857 (2020)
5. Obrosova, N.K., Shananin, A.A., Spiridonov, A.A.: A model of investment behavior of enterprise owner in an imperfect capital market. Lobachevskii J. Math. **43**(4), 1018–1031 (2022)
6. Obrosova, N.K., Shananin, A.A.: Production model in the conditions of unstable demand taking into account the influence of trading infrastructure: ergodicity and its application. Comput. Math. Math. Phys. **55**(4), 699–723 (2015). https://doi.org/10.1134/S0965542515040107
7. Alimov, D.A., Obrosova, N.K., Shananin, A.A.: Methodology for assessing the value of an enterprise in the depressed sector of economy based on solving of the Bellman equation. IFAC-PapersOnLine **51**(32), 788–792 (2018)
8. Alimov, D.A., Obrosova, N.K., Shananin, A.A.: Enterprise debts analysis using a mathematical model of production, considering the deficit of current assets. Lobachevskii J. Math. **40**(4), 385–399 (2019)
9. Shananin, A.A.: Mathematical modeling of investments in an imperfect capital market. Proc. Steklov Inst. Math. **313**(1), S175–S184 (2021)
10. Beklarian, L.A., Flerova, A.Y., Zhukova, A.A.: Optimal control methods. MIPT, Moscow (2018). (Russian)
11. Zhukova, A., Pospelov, I.: Model of optimal consumption with possibility of taking loans at random moments of time. HSE Econ. J. **22**(3), 330–361 (2018)
12. Zhukova, A., Pospelov, I.: Numerical analysis of the model of optimal consumption and borrowing with random time scale. In: Sergeyev, Y.D., Kvasov, D.E. (eds.) NUMTA 2019. LNCS, vol. 11974, pp. 255–267. Springer, Cham (2020). https://doi.org/10.1007/978-3-030-40616-5_19
13. Øksendal, B.: Stochastic Differential Equations. Springer, Heidelberg (2003). https://doi.org/10.1007/978-3-642-14394-6

Comparative Analysis of the Efficiency of Financing the State Budget Through Emissions, Taxes and Public Debt

Ivan G. Kamenev(✉) (iD)

FRC CSC of RAS, Moscow, Russia
igekam@gmail.com

Abstract. In this study, we propose an addition to the classical ISLMBP model that changes its focus. Typically, this model is used to analyze the impact of different macroeconomics policy options on macroeconomic stabilization (the return of an economy that has deviated from long-term equilibrium to a state of full resources usage). We propose an addition to this model that describes the reaction of business owners and financial investors to the level of tax burden and (more detailed) interest rates. This allows us to consider the optimization problem of replenishing the state budget from three sources: taxes, public debt and money emission. The proposed model allows us to quantitatively study the optimal proportion between these sources of financing, taking into account the specifics of the investment climate (sensitivity of the economy to the national and international interest and tax rates).

Keywords: Macroeconomics · Optimal control · Economic policy · Taxes · Emission · Public debt

1   Introduction

It is well known that the state budget can be financed by taxes, public debt and money emission. Each of these methods has its drawbacks: taxes reduce economic activity, public debt generates costs for its maintenance, and emission acts as an inflationary tax. Each of these mechanisms has been well studied separately. Surprisingly, the question of preferability for one of these mechanisms (or the proportion between all three, respectively) still remains opened. The main reason for that is that macroeconomic models are concentrated on the problems of macroeconomic policy choice (influencing an economy that has deviated from long-term equilibrium to bring it to an equilibrium state).

In this work, we propose a modification of the ISLMBP model, which allows us to formulate the problem of the optimal choice of the method of financing state social policy. Using the classical balance macroeconomic model, we obtain a

The research was supported by RSCF grant No. 22-21-00746 "Models, methods and software to support the modeling of socio-economic processes with the possibility of forecasting and scenario calculations".

universal basis, the performance of which has been verified by numerous authors. However, by adding some previously unused dependencies to it, we get the opportunity to transform it into an optimal control task.

Various economic schools (both Western and Russian: Keynesian, Neoclassic, Monetarists, Technological paradigm theory, etc.) give their answers about preferable financing method. However, none of these schools offers a macroeconomic model complete enough to take into account the impact of all three methods on economic activity and allow a quantitative analysis of the possible combination of all three methods at once. This is due to the fact that the state macroeconomic policy is considered mainly by balance-sheet macroeconomic models (based on the system of national accounts [13]), and long-term dynamics - by models of economic growth. Docking between these models is limited. Long-term and short-term dynamics of the economy are conceptualized differently.

Models of economic growth (starting with the Solow model [5], and further more modern versions: Ramsey [6], Mankiw [7], Rebelo [8] etc.) explain the dynamics of the economy through investment in factors of production (capital or, in more complex models, material capital and human capital). Accordingly, the only thing that connects these models with balance-sheet macroeconomic models is the amount of investment in production factors. And this amount is typically determined in them on the basis of some basic micro-justification, which assumes the optimal distribution of household funds between consumption and savings. However, this approach contradicts modern interpretations of macroeconomic equilibrium (based on the Keynesian-neoclassical synthesis). Thus, in balance sheet models, an increase in savings is a manifestation of the economic crisis, and there is no reason to believe that all savings will be directed to investments. Instead, they can be withdrawn from the national economy through the foreign exchange market.

The development of modern models of economic growth is associated primarily with the construction of more complex interpretations of demographic processes and human capital (as, for example, in the works of O. Galor, for example, [11]). These works are of great interest in terms of possible integration with balance sheet macroeconomic models, since they make it possible to describe changes in population size and the quality of labor resources as an endogenous processes. However, before creating such a hybrid model with endogenous labor and capital, we consider it appropriate to focus on a model with endogenous capital, which would reflect the impact of state fiscal and monetary policy on this factor in accordance with the ideas of modern macroeconomics about the dynamics of financial flows. For the description of impact of firms' decisions on financial flows and impact of financial flows on the exchange rate and investment, see, for example, [9] and [10].

Accordingly, the basic micro-justification of the amount of savings should be replaced by models that suggest the possibility of reducing the investments due to capital outflow. The basic (simplest) version of such a balance sheet model is the ISLMBP model. Note that there are more modern balance-sheet macroeconomic models that include more complex effects on the labor market, a more diverse description of financial instruments, and finer tools for external shocks description. The development of balance models in this direction is due to their focus on describing the short-term period, shocks and the anti-crisis policy of the state. Accordingly, although the use of a more complex basic balance sheet model may have its advantages, they are redundant for studying the long-term dynamics of potential GDP.

2 Model ISLMBP

The ISLMBP model is a classic macroeconomic model with a microeconomics basis, created within the scientific school of the neoclassical–Keynesian synthesis [1,2]. It describes the interaction of four macroeconomic agents (the Firm, the Household, the State and the Outside world) in three markets (Goods, Money and Foreign exchange). The agents (except State) are described by their behavioral functions: consumption, investment, and net exports, demand for the national currency, and supply of the national currency. (The full model also describes the equilibrium in the labor market, but for our purpose of the macroeconomic equilibrium, this market takes a subordinate position in relation to the goods market).

The State, on the contrary, is represented not through a function of behavior, but through a balance equation that limits the choice of the state controls: the State budget

$$Em = Debt(\tau)\, r + Tr - Ta - t\, Y - Lo - G \tag{1}$$

Here:

Y : gross domestic product (GDP)
Em : money base growth (emission)
$Debt(\tau)r$: public debt service costs charged on public debt at an interest rate.

Note that all money cost interest rates r are considered the same throughout the model since it is assumed that differences in rates on different financial instruments in highly competitive financial markets are fully explained by differences in the level of risks.

Also note that formally all variables of the balance-sheet model (not including sensitivity factors) are formally dynamic (take on different values at different time moments τ). In order to simplify the notation, here and below, only variables included in differential equations are marked as dynamic (see Sect. 4). The remaining variables are fully calculated on the basis of the macroeconomic balances (see IS, LM and BP).

Tr: social transfers to the Household
Ta: Autonomous (GDP-independent) taxes (such as property taxes)
t in $(0,1)$: level of tax burden on economic activity (VAT, profit taxes, labor taxes, consumption taxes, etc.)
Lo: additional government borrowing (government debt growth)
G: government spending.

Let us emphasize that we equally classify as emission all the forms of money base generations. In particular, the direct printing of money by the Treasury, the repurchase of state loan bonds by the Central Bank, and other forms of "quantitative easing" (the repurchase of securities and stocks by the Central Bank).

To simplify things, instead of full descriptions of agents, we present the final balance equations of their interaction in the markets (known as IS, LM, and BP):

$$IS : Y = \frac{(Ca + Ia]; K(\tau) - dr\, r + G + NXa + MPC(Tr - Ta))}{(MPC\, t - MPC + MPI + 1)} \quad (2)$$

Here:

Ca: autonomous consumption (living wage, etc.)
$IaK(\tau) - drr$: investments aimed at expanding/replenishing capital (fixed assets), reduced by the opportunity cost of financial capital
r: risk-free investment rate of return (on deposits, bonds, etc., associated with the key rate of the Central Bank)NXa: autonomous net export (difference between export and import)
MPC in $(0,1)$: marginal propensity to consumption (part of household income growth directed towards consumption rather than saving)
MPI: marginal propensity to import, part of expenses spent on imported rather than domestic goods.

$$LM : Y = \frac{(cr + 1)}{(cr + rr)} \frac{MB(\tau)V}{P} + h\, r \quad (3)$$

Here:

MB: monetary base (simply speaking, government-issued national currency)
cr: deposit ratio (percentage of funds circulating as cash outside the banking system)
rr: reserve ratio (percentage of banks' assets not issued as loans)
h: sensitivity of the investment, "speculative" demand for money according to the interest rate
V: velocity of money circulation (institutional characteristic associated primarily with the level of development of payment systems and peoples' expectations)
P: prices level in the economy (phase variable)

$$BP : Y = \frac{NXa + Fa + f(r - re)}{MPI} \quad (4)$$

Here:

Fa: autonomous capital export/import
f: sensitivity of capital export to the difference between the national interest rate r and the rate re on international markets (The ISLMBP model assumes that the government cannot of maintain national currency overvalued or undervalued for a long time).

3 Model Extension: ISLMBPFI

The ISLMBP model allows various interpretations (see [3]). We use a small open economy interpretation, in which the rate of interest in the outside world does not depend on events in the national economy (although our model can be adapted to the case of a large open economy).

The classical ISLMBP model considers the export/import of capital as a process that depends exclusively on the interest rate. This shortcoming follows from the microeconomics basis of the model, which did not consider the question: where do the funds go to the foreign exchange market come from (if the balance of financial capital is negative), and where do the funds go in the economy (if the balance of capital is positive). We propose a new interpretation of the flow of financial capital.

Consider briefly the choice of the Firm's owner in the national economy. (In terms of microeconomics, the owner is a special household whose income consists mainly of dividends). He accumulates financial resources (by the rate r), invests them in the renewal and development of production (investment) and expects to receive a profit. Note that professional financiers are also owners (of financial sector firms). This also applies to the distribution of profits: the owner invests part of the profits back into his own business (in this case, the alternative cost is the same interest rate r at which he can place them on the financial market). The lower the interest rate of raising capital is, the greater profit owners can count on; and the higher the level of tax burden on economic activity t is, the lower the profit will be. Thus, the amount of investment in the national economy is inversely proportional to the interest rate r and the tax rate t.

At the same time, in an open economy, the owner has a choice: to invest in the national economy, or withdraw funds abroad, where conditions for investment are better. The higher the interest rate abroad re are, and the higher the foreign taxes te are, the lower desire of the owner to withdraw funds abroad is.

Of course, the search for investment places in foreign economies is associated with increased transaction costs, so only a certain part of national investors will withdraw money abroad: however, the greater the gap in conditions is, the larger will be the proportion.

$$I = K(\tau)\, Ia - dr\, r - dt\, t - fr\, re + ft\, te \qquad (5)$$

Here: dr, dt, fr and ft: sensitivity of investment to the national interest rate r and the rate re on international markets, national tax burden t and the burden te on international markets.

In turn, in the foreign exchange market, there is an interaction of domestic and foreign owners. Both domestic and foreign owners buy more national currency if the interest rate on it is higher than the world one, and the tax burden is lower (and vice versa).

$$0 = NXa - MPI\, Y + Fa + rf\, (r - re) + tf\, (te - t) \qquad (6)$$

Here: rf and tf: sensitivity of capital export to the difference between the national interest rate r and the rate re on international markets, national tax burden t and the burden te on international markets.

.These additions, describing the relationship between investment and the movement of financial capital, significantly change the overall equilibrium in the ISLMBP model, but, more importantly, allow it to reflect the trade-off between different methods of the state budget financing. Without them, tax financing of the state budget did not have significant disadvantages, due to the nature of multipliers: simultaneous proportional increase in taxes and government spending should increase GDP. After adding this balances, tax financing of social policy also has negative consequences for GDP growth rates (similar to how emission financing can lead to inflation, and public debt generates budget spending).

Thus, the equilibrium conditions in the markets of goods, money and foreign currency take will be:

$$Y = \frac{Ca + IaK(\tau) - dr\,r - fr\,re - dt\,t - ft\,te + G + NXa - MPC(Ta - Tr)}{MPC\,t - MPC + MPI + 1}$$

(7)

$$Y = \frac{(cr + 1)MB(\tau)V}{(cr + rr)P} + h\,r$$

(8)

$$Y = \frac{NXa + Fa + rf(r - re) + tf(te - t)}{MPI}$$

(9)

This system of three equations with three unknowns (P, r, Y) is unequivocally solved. A significant feature of the resulting solution is that the key rate of the Central Bank is not only not a control, but is in fact a function of the level of the tax burden. The Central Bank is forced to set it at the level calculated in the model to avoid imbalance in the balance of payments. This means that in a small open economy (with high taxes), pressure from capital outflow prevents the Central Bank from targeting inflation (changes in the prices level P).

4 · The Statement of the State's Problem of Optimal Control

The State controls a significant number of model variables, although some of them are considered ineligible for change. The government controls Tr, Ta, t and G (fiscal policy). The central bank controls rr (monetary policy). The Central Bank and the Government jointly control the amount of public debt Lo. At the same time, a change in Ta and rr is considered to be dangerous (they are not limited in this model, but any changes to them require verification of indirect consequences for the state budget using auxiliary models).

We emphasize that this statement corresponds to the logic of choosing a state that successfully copes with the implementation of a counter-cyclical/stabilization policy. That is why the monetary policy seems to be redundant (the Central Bank just needs to set r in accordance with the model equilibrium). The main issue of public policy is:

– considering the volume of social obligations G and Tr given (politically), and Ta and rr unchangeble
– choose a financing method: the optimal volume Lo and t (as well as the volume of money emission Em calculated from this).

As in the Solow model, the dynamics of capital in the model (given State controls) can be expressed analytically. This makes it possible to study the objective function at least by simulation methods (and potentially to obtain an analytical solution on the optimal value of the State controls).

To answer this question, it is necessary to compose a functional (optimal choice criteria). In the macroeconomic literature, three such criteria are most common: the capital, the growth rate of (potential) GDP and the ratio of GDP to public debt. In both cases, an auxiliary model of economic growth is needed. In the macroeconomic literature, such models are based, as a rule, on the classical Solow model [5], using the Cobb-Douglas production function [4]. It describes economic growth as a process of capital accumulation in the economy. The Solow-type model explains it by a constant rate of savings converted into investments. More sophisticated models take savings out of macroeconomic equilibrium ([6, 12]. We use the same approach, but take into account the impact on investment of what we call the investment climate: the sensitivity of owners to global and national interest rates, as well as the global and national tax burden.

$$Ylr(\tau) = bK(\tau)^a L^{1-a} \qquad (10)$$

$$\frac{\mathrm{d}K(\tau)}{\mathrm{d}\tau}) = I - K(\tau)\,am \qquad (11)$$

$$\frac{\mathrm{d}MB(\tau)}{\mathrm{d}\tau}) = Em \qquad (12)$$

$$\frac{\mathrm{d}Debt(\tau)}{\mathrm{d}\tau}) = Lo \qquad (13)$$

Note that the capital intensity of capital (the amount of investment required to form a unit of capital) can be specified both in the capital transfer equation or in the production function.

$Ylr(\tau)$: the potential GDP (obtained from the full usage of the resources of the national economy).

a, b: calibration constants reflecting scale effect and production technology.

Assuming (for simplicity) the working-age population to be permanent and independent of state social policy, we obtain three phase (stock) variables: capital, money base, and public debt. Potential GDP can be calculated through the change in capital:

$$F(Lo, t) = \int_{\tau 0}^{\infty} \frac{Ylr(K(\tau))}{Debt(Lo)} \mathrm{d}\tau \qquad (14)$$

In turn, the change in capital is uniquely determined by the controls Lo and t.

$$K(\tau) = K0 \, e^{\frac{\tau((Ia-am)((t-1)MPC+MPI+1) \, rf - am \, dr \, MPI)}{((t-1)MPC+MPI+1)rf+MPI \, dr}}$$

$$- \frac{(dr \, re + dt \, t + fr \, re - ft \, te)((t-1)MPC + MPI + 1) \, rf}{(am - Ia)((t-1) \, MPC + MPI + 1) \, rf + am \, dr \, MPI}$$

$$+ \frac{dr((Ta - Tr) \, MPI + (t-1)(te \, tf - t \, tf + Fa + NXa)) \, MPC}{(am - Ia)((t-1)MPC + MPI + 1)rf + am \, dr \, MPI}$$

$$+ \frac{dr(te \, tf - t \, tf - Ca + Fa - G) \, MPI + tf \, te - tf \, t + NXa + Fa)}{(am - Ia)((t-1)MPC + MPI + 1)rf + am \, dr MPI} \quad (15)$$

$$F(Lo, t) = \int_{\tau 0}^{\infty} \frac{Ylr(t, Lo)}{Debt(Lo)} d\tau \quad (16)$$

This functional can be discounted, but we consider its relevance questionable for reasons beyond the scope of this study.

Since the value of potential GDP in the model is uniquely calculated through the value of capital, then by deriving the differential equation that describes the dynamics of capital (with the values of controls chosen by the state), we also obtain the explicit form of the functional $F(t, Lo)$. (The substitution (15) into (10) and then into (16) is not reproduced in the article due to being cumbersome).

The maximization of this functional is analytically possible, but difficult due to the difference in the powers of the numerator and denominator. Note that the problem can be easily solved analytically if we represent the functional as

$$F(Lo, t) = \int_{\tau 0}^{\infty} \frac{K(\tau)}{Debt(Lo)} d\tau \quad (17)$$

However, such a replacement is unacceptable, because diminishing returns on capital means that the ratio of GDP to debt may decrease as the ratio of capital to debt increases.

Note that long-term GDP Ylr, strictly speaking, is not identical to short-term GDP Y, since the ISLMBP model allows the economy to deviate from the optimal level due to supply and demand shocks. However, as in the neoclassical model of economic growth, we consider the economy in a stable state, not in a state of shock. Then the mutual correspondence of short-term and long-term GDP is ensured by the calibration of the Cobb-Douglas function (coefficients and capital K itself, since it is an unobserved characteristic).

5 Model Trajectories' Features

Let us present the results of a preliminary analysis of the model dynamics associated with the policies of some developed countries. Consider three cases:

1. financing mainly through public debt
   (Japan; USA considering some specifics)

2. financing mainly through emission (Venezuela)
3. financing mainly through taxes (France).

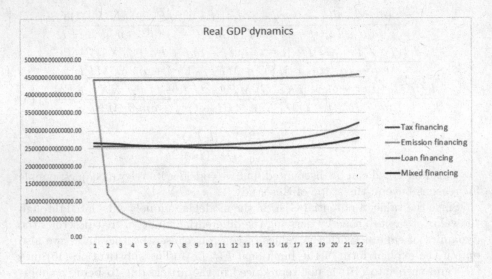

Fig. 1. Real GDP dynamics in quantitative example

Let us give an example of the dynamics of the model with these three types of policies from the same initial state of the national economy (see Fig. 1 and Fig. 2). Key model parameters are: $MPC = 0.5$, $MPI = 1$ $V = 2$, $h = 100$, $cr = 1.6$, $rr = 0.05$, $re = 0.02$, $te = 0.3$. This analysis was made by a discrete delta version of our model. A fixed amount of government spending $(G + Tr)$ is financed here by each of the three ways $(t = 0.38$, or $Lo = G + Ta$, or $E = G + Ta)$, as well as a mixed option (some tax cuts, a small government debt and moderate emission at a level that maintains the prices level). This is just an example, which, in particular, does not take into account the deflationary trap. We should not base on it any meaningful conclusions about the advantages of one or another method of financing. This will become possible only after the identification of the model according to the macroeconomic statistics of several countries.

However, this example (with parameters broadly reminiscent of the Russian economy) makes it possible to make sure that the patterns in the model are quite realistic, and the trajectory of economic development changes greatly while choosing different financing methods. For example, it can be seen that inflationary financing leads to a short-term rise in GDP followed by a sharp fall, while debt financing allows maintaining GDP at an inflated level at the cost of losing growth rates. Both observations are consistent with generally accepted ideas about macroeconomic dynamics.

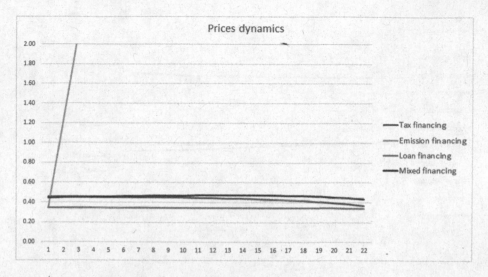

Fig. 2. Inflation dynamics in quantitative example

At the same time, it can be shown that the dynamics observed in the model changes strongly when the fundamental parameters change. For example, if the sensitivity of investment and financial capital to the national tax rate is sufficiently high (see Fig. 3), then tax financing of social policy becomes less productive, in contrast to debt (which is typical for the United States, thou formally Unated States should not be considered as a small open economy). And vice versa, in countries with low inequality and a high quality of life (which contribute to reduced tax sensitivity), the level of the tax burden can be significantly higher than the global one without serious consequences for the balance of payments and investment activity ("Scandinavian socialism"). Moreover, the model allows us to describe the "resource curse": the state of the economy with large autonomous exports (raw resources) and a poor institutional setting. An attempt to finance major social programs of the state through taxes leads in this case to the decline of domestic investment and stagnation in the economy, and low taxes lead to problems with the balance of payments ("Dutch disease").

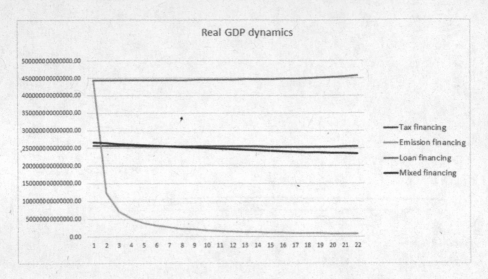

Fig. 3. Inflation dynamics in quantitative example for higher taxes sensibility

6 Conclusions and Perspectives

The proposed ISLMBP model with the addition FI, which describes the impact of taxes and interest rates on investment and financial capital flow, is designed to compare different ways of financing the state budget: money emission, taxes, and loans. It allows us to explore different combinations of these sources and predict the reaction of owners/investors to them. Additional coefficients introduced into the model make it possible to consider it as an optimal control model. All ways of the budget financing can be modeled in it as having maxima in the amount of funds raised and GDP value.

The analytical solution of the problem of optimizing the growth rate of GDP, capital or debt-to-GDP ratio is mathematically complex, but can be researched by simulation methods. The model can be adjusted for countries with different investment climates (sensitivity of investors and financial capital to interest rates and taxes). The next stage in the development of the model may be its adaptation to various monetary policies (primarily, to the policy of accumulating gold and foreign exchange reserves). After that, it will be possible to identify the model on the macroeconomic statistics of different countries, including Russia.

Another interesting, but mathematically difficult, possibility for the development of the model is related to the decomposition of the marginal propensity to import into different motives: technological and consumer exports and imports. This will make it possible to abandon the rigid identification of autonomous exports according to macro statistics, considering its changes during the structural restructuring of the economy.

References

1. Mundell, R.A.: Capital mobility and stabilization policy under fixed and flexible exchange rates. Can. J. Econ. Polit. Sci. **29**(4), 475–485 (1963). https://doi.org/10.2307/139336
2. Fleming, J.: Marcus domestic financial policies under fixed and floating exchange rates. IMF Staff. Pap. **9**, 369–379 (1962)
3. Young, W., Darity, W.: IS-LM-BP: an inquest. Hist. Polit. Econ. **5**(36), 127–164 (2004)
4. Cobb, C.W., Douglas, P.H.: A theory of production. Am. Econ. Rev. **18**, 139–165 (1928)
5. Solow, R.M.: A contribution to the theory of economic growth. Quart. J. Econ. **70**(1), 65–94 (1956)
6. Ramsey, F.P.: A mathematical theory of saving. Econ. J. **38**(152), 543–559 (1928). https://doi.org/10.2307/2224098
7. Mankiw, N.G., Romer, D., Weil, D.N.: A contribution to the empirics of economic growth. Q. J. Econ. **2**(107), 407–437 (1992)
8. Rebelo, S.: Long-run policy analysis and long-run growth. J. Polit. Econ. **3**(99), 500–521 (1991)
9. Itskhoki, O., Mukhin, D.: Exchange rate disconnect in general equilibrium. J. Polit. Econ. **8**(129), 2183–2232 (2021)
10. Gaubert, C., Itskhoki, O., Vogler, M.: Government policies in a granular global economy. J. Monet. Econ. **121**, 95–112 (2021)
11. Galor, O., Weil, D.N.: Population, technology, and growth: from Malthusian stagnation to the demographic transition and beyond. Am. Econ. Rev. **4**(90), 806–828 (2000)
12. Koopmans, T.C.: On the concept of optimal economic growth. The Economic Approach to Development Planning, pp. 225–287. Rand McNally, Chicago (1965)
13. System of National Accounts 2008. Unated Nations. https://unstats.un.org/unsd/nationalaccount/sna2008.asp

Applications

Construction of Optimal Feedback for Zooplankton Diel Vertical Migration

O. Kuzenkov$^{(\boxtimes)}$ (iD) and D. Perov (iD)

Lobachevsky State University, Gagarin Avenue 23, 603950 Nizhny Novgorod, Russia
kuzenkov_o@mail.ru,diper1998@yandex.ru
http://www.unn.ru

Abstract. We consider the optimization problem of forming the fittest strategy for zooplankton diel vertical migration. This strategy should maximize the fitness function reflecting the average specific rate of population reproduction. We solve this problem using feedback between the current environmental state and the organism's local movement. Such feedback reflects the ability of living organisms to adapt to changing habitat conditions. We construct the feedback on the base of the neural network. Its input is the values of environmental factors at a given point and a given time; its output is the corresponding local displacement of zooplankton. The initial optimization problem is reduced to the optimization of the feedback settings or to the optimal choice of the neural network weights. To train the neural network, we apply the new evolution method of stochastic global optimization: Survival of the Fittest by Differential Evolution (SoFDE), based on the Survival of the Fittest algorithm and Differential Evolution. It was shown that this approach permits to form the optimal behavioral strategy for different environmental conditions.

Keywords: Feedback · Global optimization · Survival of the Fittest algorithm · Differential evolution · Neural network · Zooplankton · Diel vertical migration

1 Introduction

In population biology, we are generally interested in a certain behavioral strategy (or a life trait) that remains in the community as a result of competition [30]. The knowledge of this strategy allows us to understand and predict outcomes of long-term biological evolution. Darwin's idea of "survival of the fittest" suggests that the strategy persisting in the population for a relatively long period of time is the one that should have the highest fitness [4,12]. In the simplest cases, the evolutionary fitness is proportional to the average specific energy gain, which is used for reproduction of the corresponding subpopulation [14,22,28]. Biologically more justified to construct feedback that would connect the current environmental state and the organism's local behavior. Such feedback provides the ability of living organisms to adapt to changing habitat conditions. It forms different fittest strategies for different biotic/abiotic external factors. In this case, the initial problem is reduced to the choice of the optimal feedback settings.

This problem arises, in particular, in modeling diel vertical migration (DVM) of zooplankton. The phenomenon of diel vertical movements of aquatic organisms was discovered two hundred years ago [7] and it consists of regular ascending and descending of zooplankton in the water column. Diel vertical movements of zooplankton play an important role in the dynamics of the organic matter of the ocean. They are the most significant synchronous movement of biomass on Earth [21] and can potentially have an influence on the planet's climate [1,6,9,18]. Identifying the causes and mechanisms of DVM is an important problem in modern ecology. The effect of diel vertical movements of zooplankton has been studied by many scientists both empirically and theoretically [2,10,15,16,34]. DVM of zooplankton is considered as the result of adaptation to the dynamical environmental factors during biological evolution [11,17,24], however, many aspects of this phenomenon are still poorly understood. This is due to the wide variety of patterns of DVM observed in nature. Nature observations show that the strategy of DVM changes with the season chaining of environmental conditions. It is known that some species of zooplankton carry out detectable vertical movements, but for others, this is not typical [7]. In addition, certain species of zooplankton make detectable movements only when they reach a certain age. The complexity of the problem of modeling DVM is aggravated by the inevitable random spread of the above factors and stochasticity of different nature.

The purpose of this work is to construct optimal feedback between the current environmental state and the organism's local behavior on the base of fitness maximization. This feedback permits to form adaptive zooplankton DVM strategies. To solve this problem, we use a neural network to approximate the optimal feedback function. Neural networks, as universal function approximators, can be effectively applied to the problem of finding feedback [40], and it is worth noting that the neural network is the mathematical model, which is based on the principle of functioning of nerve cells networks of living organisms. Also, we use global optimization methods for training neural networks. Currently there are many global optimization methods [36,37,39,41]. Here we use evolutionary algorithms because they are based on the ideas of mutation, crossover and selection and are suitable for modeling the phenomena of biological evolution. The following evolution methods of stochastic global optimization are proposed: the Differential Evolution [38] which can be considered as the suitable algorithm for training the neural network [3] and the novel method - Survival of the Fittest by Differential Evolution (SoFDE), based on the Survival of the Fittest algorithm proposed in [23,29]. An important advantage of SoFA is that one can rigorously guarantee its convergence for an extensive class of objective functions [29]. Another significant advantage of SoFA is that this framework can efficiently work in higher dimensional spaces, spaces of increasing dimension or even infinite-dimension Hilbert spaces. This algorithm shows the higher convergence rate than several other evolution methods (Evolutionary Strategy with Cauchy distribution [35], Controlled

Random Search with local mutation [20,32,33] and Multi Level Single-Linkage) for some important classes of optimization problems in the spaces of a high dimension. It was tested in a class of relevant objective functions and was successfully applied for investigations of physical and biological models [23,29]. The SoFDE method was used for the first time. In this method we have retained the advantages of SoFA and improved the convergence rate using Differential Evolution. Then we compared the efficiency of the SoFDE algorithm with the Differential Evolution and the original SoFA in neural network training.

This work is motivated by the need to reveal the mechanisms of zooplankton adaptation by modeling optimal feedback, which ensures the formation of the most fitness strategies. The contribution of this paper is the construction of optimal feedback in the problem of modeling zooplankton DVM using neural networks and evolutionary algorithms and the demonstration of its effectiveness on field observations. The paper is organized as follows. At the beginning, we describe the environment for the zooplankton DVM and introduce the fitness function. Meanwhile, we consider the problem of finding the best fit zooplankton DVM strategy and discretize the problem. Then we describe the neural network architecture and present the evolutionary algorithm SoFDE for training the neural network. We construct the optimal feedback between zooplankton movements and the environment. Finally, we form the migration strategy using the constructed optimal feedback and test it under different environmental conditions.

2   Materials and Methods

2.1   The Problem Statement

Empirical data show that the movement of plankton is determined by various environmental factors: spatial distributions across the depth x of food (phytoplankton) $F(x)$, predator (fish) density $P_x(x)$, distribution of adverse factors $D(x)$, such as temperature, radiation level etc., the overall predator activity $P_t(t)$ during the day, etc. [7,10,27]. All of these factors can be considered as mathematical functions of the vertical coordinate x (measured in meters) or the time of day t (measured in hours). The activity of predators depends on daylight; the function $P_t(t)$ is periodic with the period 24 h.

Here we use biological relevant parameters determining the DVM of a dominant zooplankton herbivorous species in the Black Sea [27]. We selected these parameters so as to find an optimal migration strategy that would satisfy the obtained observations. We use the following parametrization of F, P_x, P_t and D:

$$F(x) = \frac{\tanh(\xi_1(x - c_1)) + 1}{2}, \tag{1}$$

$$P_x(x) = \tanh(\xi_2(x - c_2)), \tag{2}$$

$$P_t(t) = \cos\left(2\pi(\frac{t}{24} + \frac{1}{2})\right) + 1, \tag{3}$$

$$P(x,t) = \frac{1}{2}P_x(x)P_t(t), \tag{4}$$

$$D(x) = \frac{\xi_3 \eta_1^{x-c_3} + \xi_4 \eta_2^{-(x-c_4)}}{2}. \tag{5}$$

Here c_1, c_2, c_3 and c_4 are some characteristic depths, ξ_1, ξ_2, ξ_3, ξ_4 and η_1, η_2 are some constants. Note that this parameterization of the vertical distribution of food $F(x)$ is justified by empirical observation [27], likewise this concerns the vertical distribution of visual predation load $P_x(x)$ [10]. The parameterization of the natural mortality $D(x)$ is in fact an approximation of the sum of the two sigmoid functions used in [27] which has empirical justification. The values of parameter functions F, P, D are taken close to those considered in [27].

Consider x as the vertical coordinate of the position of zooplankton; t is the time of day varying from 0 to 24; $\nu = x(t)$ is the hereditary strategy of plankton behavior - a continuous differentiable function on the segment $[0; 24]$, satisfying condition $x(0) = x(24)$. In the simplest case, the fitness function of the strategy ν has the following form [28]:

$$J(\nu) = \int_0^{24} (\alpha F(x) - \beta P(x,t) - \gamma D(x) - \delta E(x')) \, dt. \tag{6}$$

Here we assume that the energetic cost $E(x')$ of the vertical movement is proportional to the square of its velocity $x'(t)$

$$E(x') = \xi_5 (x'_t(t))^2. \tag{7}$$

The weighting coefficients $\alpha, \gamma, \beta, \delta$ quantify the relative contribution of the corresponding environmental factors.

We can find the strategy with the highest fitness by methods of the calculus of variation. But it is much more interesting and more promising to form the best strategy based on the perceived current value of environmental factors at a given position of zooplankton at a given time. In other words, our problem is to construct feedback between the environmental state and the zooplankton movement, which leads to the optimal strategy during the day. Such feedback allows zooplankton to adapt to changes in the environment. We solve the problem with the help of a neural network. We will train the network for fixed environmental functions, and then we will consider its work for different functions.

Here we use piecewise linear approximations of zooplankton movement strategies. We divide the day into n equal intervals (t_i, t_{i+h}), where $h = 24/n$, $t_0 = 0$, $t_i = t_0 + ih$, and consider that the depth of zooplankton immersion varies linearly at each interval. In this case, the zooplankton strategy is determined by the positions $x(t_i) = x_i$ of the zooplankton at times t_i or by the initial position $x_0 = x(0)$ and displacements $\Delta x_i = x_i - x_{i-1}$. We consider the displacements to be of the following form $\Delta x_i = \upsilon \sigma_i$ where υ is a fixed constant and coefficients σ_i can have one of three values $-1, 0, 1$. Then the behavior strategy and the corresponding fitness value are determined by the initial position x_0 and the set of coefficients σ_i : $J(\nu) = J(\nu(x_0, \sigma_1, \ldots, \sigma_n))$. It is necessary to find the optimal values of σ_i.

We solve this problem using feedback – we construct the dependence of the displacement coefficient σ_i on the values of environmental factors $F(x_{i-1})$, $P(x_{i-1}, t_{i-1})$, $\Delta P(x_{i-1}, t_{i-1}) \equiv P(x_{i-2}, t_{i-2}) - P(x_{i-1}, t_{i-1})$, $D(x_{i-1})$ at a given point x_{i-1} and a given time t_{i-1} in order to provide the maximum of the fitness function $J(\nu)$. In other words, our goal is to find the function $\sigma : \sigma_i = \sigma(F(x_{i-1}), P(x_{i-1}, t_{i-1}), \Delta P(x_{i-1}, t_{i-1}), D(x_{i-1}))$ to maximize the functional: $J(\nu(x_0, \sigma(F(x_0), P(x_0, t_0), \Delta P(x_0, t_0), D(x_0))), \ldots, \sigma(F(x_{n-1}), P(x_{n-1}, t_{n-1}), \Delta P(x_{n-1}, t_{n-1}), D(x_{n-1}))))$. To construct this dependence σ, we use the neural network. In this case, this dependence is defined by the set N of the network weights: $\sigma_i = \sigma(F(x_{i-1}), P(x_{i-1}, t_{i-1}), \Delta P(x_{i-1}, t_{i-1}), D(x_{i-1}), N)$. Thus, the initial problem is reduced to finding the optimal setting N of the network.

It is worth noting that the original problem can be reduced to the well-known class of optimal control problems with feedback [25]. It can also be considered as the inverse problem: we need to find the function $x(t)$ by given functions (1–5) where the functional (6) is maximized.

2.2   Construction of Optimal Feedback

To solve the problem, we have built a four-layer neural network. The architecture of the artificial neural network is shown in Fig. 1. Such the neural network architecture is justified by systematic experiments and works [19, 26].

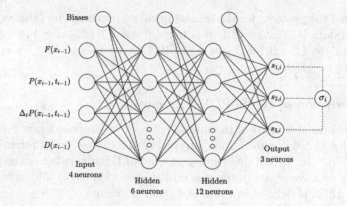

Fig. 1. The architecture of the artificial neural network.

The first input layer contains four fully connected neurons, the second and third hidden layers contain 6 and 12 fully connected neurons respectively and the fourth layer contains 3 neurons; w_i is the weight of i-th connection – the amplification coefficient of the signal passing through the i-th connection, $|w_i| < w_{max}$; b_j is the bias corresponding to j-th neuron of the second - fourth layers, $|b_j| < b_{max}$. The activation function for every neuron of the second - fourth layer has the form of the sigmoid:

$$f(y) = \frac{1}{1 + e^{-y}}. \tag{8}$$

Thus, the network is described by a set N of weights w_i, $i = 1,132$, and biases b_j, $j = 1,21$. Input is values of four environmental factors of the current zooplankton position. The output determines displacement coefficients in accordance with the following rule:

$$I(s_{1,i}, s_{2,i}, s_{3,i}) = \begin{cases} -1, \text{if } \max(s_{1,i}, s_{2,i}, s_{3,i}) = s_{1,i}; \\ 0, \text{if } \max(s_{1,i}, s_{2,i}, s_{3,i}) = s_{2,i}; \\ 1, \text{if } \max(s_{1,i}, s_{2,i}, s_{3,i}) = s_{3,i}. \end{cases} \tag{9}$$

$$\sigma_i = I(s_{1,i}, s_{2,i}, s_{3,i}). \tag{10}$$

We use the following characteristics of the neuron network: $w_{max} = b_{max} = 100$. Knowing the current zooplankton position $x(t_{i-1})$, we can calculate the corresponding environmental characteristics, then pass this information through the neural network and calculate the displacement and new zooplankton position $x(t_i)$. Thus, we can obtain the zooplankton daily trajectory $\nu(N) = x(t, N)$ corresponding to a given neural network N by repeating this procedure n times. Then we trained the constructed neural network - selected the values of weights and biases that maximize fitness:

$$J(\nu(N^*)) = \max_{\substack{N \\ |w_i| \leq w_{max}, \\ |b_j| \leq b_{max}}} (J(\nu(N))). \tag{11}$$

We solved the problem for the following values of parameters that correspond to observed data [8,27]: $\alpha = 0.4$, $\beta = 0.46$, $\gamma = \delta = 0.025$, $n = 1440$, $\xi_1 = 0.02$, $\xi_2 = 0.025$, $\xi_3 = \xi_4 = 0.105$, $\xi_5 = 0.27 \cdot 10^{-3}$, $\eta_1 = \eta_2 = 1.15$ $c_1 = c_2 = -100$, $c_3 = -10$, $c_4 = -120$, $\upsilon = 1$, $x_0 = -15$.

2.3 Description of the SoFDE Framework

To find the optimal weights of the neural network we propose a new evolutionary algorithm for global optimization: Survival of the Fittest by Differential Evolution (SoFDE). The formal description of the SoFDE algorithm is the following.

Suppose we have some continuous positive function $J(N)$ (objective function or fitness) which is defined in the rectangular domain:

$$\Pi = \{N = (w_1, \ldots, w_m) : W_{min} \leq w_j \leq W_{max}, j = \overline{1,m}\}. \tag{12}$$

Here W_{min} and W_{max} are some constants. Assuming that $J(N)$ has a unique vector of global maximum (denoted by N^*) in Π.

A population P_g consists of NP agents or vectors: $P_g = (N_{1,g}, \ldots, N_{NP,g})$, where g denotes the generation index, $g = \overline{1, G_{max}}$.
Each vector $N_{i,g} = \{w_{i,1,g}, w_{i,2,g}, \ldots, w_{i,m,g}\}$, $i = \overline{1, NP}$ consists of m variables.

0. *Initialization.* Before the evolutionary process begins, the population is randomly initialized, each vector receives randomly generated uniformly distributed values between the lower and upper bounds of its components: $w_{i,j,1} = rand(W_{min}, W_{max})$.

Then the population P_g evolves and directs vectors in the search space to the global optimum. At the end of the evolutionary process, a vector with the maximum value of the fitness function is returned as the final solution. During each generation, evolutionary algorithms use three operations for each agent, its Mutation, Crossover, and Selection.

1. *Mutation.* The mutant vector $\widetilde{N}_{i,g+1}$ is created using the mutation operator, which is described below:

   (a) Each vector in the population is assigned once per generation for NP iterations the probability of participation for further mutation as a reference vector:

   $$\frac{J^{\psi_g}(N_{i,g})}{J^{\psi_g}(N_{1,g}) + \ldots + J^{\psi_g}(N_{NP,g})}. \tag{13}$$

   Here ψ_g is a parameter of the method, an infinitely increasing sequence, depending on the generation, regulating the rate of convergence. Given the probabilities found, the reference vector $N_{r,g}, r \in \{1, \ldots NP\}$ and its corresponding coordinates $w_{r,j,g}$ are randomly selected from the population.

   (b) The mutant vector $\widetilde{N}_{i,g+1}$ is created randomly, the components of which $\widetilde{w}_{i,j,g+1}$ take values on the segment $[W_{min}, W_{max}]$ with probability density:

   $$\frac{A_{i,r,j,g}\varepsilon_{i,g}}{\varepsilon_{i,g}^2 + (\widetilde{w}_{i,j,g+1} - w_{r,j,g})^2}. \tag{14}$$

   Here $\varepsilon_{i,g}$ is a sequence decreasing to zero, and $A_{i,r,j,g}$ is the normalizing probability density constant for the segment $[W_{min}, W_{max}]$:

   $$A_{i,r,j,g} = (arctan(\frac{w_{max} - w_{r,j,g}}{\varepsilon_{i,g}}) - arctan(\frac{w_{min} - w_{r,j,g}}{\varepsilon_{i,g}}))^{-1}. \tag{15}$$

   In other words, a mutant vector is obtained with the following mutation of its components:

   $$\widetilde{w}_{i,j,g+1} = w_{r,j,g} + \varepsilon_{i,g}tan((rand(0,1) - \frac{1}{2})A_{i,r,j,g}^{-1}). \tag{16}$$

   For example, let $w_{r,j,g} = 0$, $A_{i,r,j,g} = \pi^{-1}$ for $j = \overline{1,m}$ and $\varepsilon_{i,g} = 1$, then the components of mutant vector $\widetilde{N}_{i,g+1}$ have the standard Cauchy distribution:

   $$\widetilde{w}_{i,j,g+1} = tan((rand(0,1) - \frac{1}{2})\pi). \tag{17}$$

   This mutation operation corresponds to the SoFA algorithm, but with some modifications, it uses the Cauchy distribution instead of the Gaussian distribution.

2. *Crossover.* The population of test vectors is divided into two parts: $\bar{N}_{q,g+1}$, $q = \overline{1, Q_{max}}$ and $\bar{N}_{l,g+1}$, $l = \overline{Q_{max}+1, NP}$. The crossover operator is not applied for the first part $\bar{N}_{q,g+1}$, $q = \overline{1, Q_{max}}$, where $Q_{max} < NP$. This means that the test vectors of the first part will receive a mutation for all

their components: $\bar{w}_{q,j,g+1} = \widetilde{w}_{q,j,g+1}$, where $j = \overline{1,m}$ and $g = \overline{1, G_{max}}$. For the second part the created mutant vector $\widetilde{N}_{l,g+1}$ participates in the formation of the test vector $\bar{N}_{l,g+1}$ as follows:

$$\bar{w}_{l,j,g+1} = \widetilde{w}_{l,j,g+1}, \text{ if } rand(0,1) \leq CR_{l,g} \text{ or } j = j_r, \text{ else } w_{l,j,g}, \qquad (18)$$

for $l = \overline{Q_{max}+1, NP}$ and $j = \overline{1,m}$. Crossover parameter $CR_{l,g} \in [0,1]$ represents the probability of selecting components for the test vector from the mutant vector. The randomly selected index $j_r \in \{1, 2, \ldots, m\}$ is responsible for ensuring that the test vector contains at least one component from the mutant vector. If the component was not selected from the mutant vector, then it is taken from the parent vector $N_{i,g}$. We recommend setting the size of the first part Q_{max} as 10% of NP. This crossover operation corresponds to the DE algorithm, but with modifications, it does not apply to some vectors.

3. *Selection*. After the crossover operation, the test vector is evaluated – the fitness function $J(\bar{N}_{i,g+1})$ is calculated, then its value is compared with the corresponding value from the population $J(N_{i,g})$. The best vector will remain in the next generation:

$$N_{i,g+1} = \bar{N}_{i,g+1}, \text{ if } J(N_{i,g}) \leq J(\bar{N}_{i,g+1}), \text{ else } N_{i,g}. \qquad (19)$$

This selection operation corresponds to the DE algorithm.

We introduced the algorithm SoFDE that uses basic steps, such as mutation, crossover and selection from DE, but with the modified mutation and crossover operator, which carry the ideas of SoFA.

The stopping criteria can be expressed by the maximum number of fitness function calculations, a time limit, or reaching the required accuracy. Assume that the calculation of a multidimensional function spends much more time than the rest of the algorithm's work. Therefore, let's introduce the maximum number of iterations – calculations of the fitness function K_{max} and a variable that tracks the current number of calculations of the fitness function $k = (g-1)NP + i$, $g = \overline{1, G_{max}}$, $i = \overline{1, NP}$. The algorithm finishes its work if $k > K_{max}$.

The SoFDE algorithm has several configurable hyperparameters: NP, $CR_{i,g}$, $\varepsilon_{i,g}$, which can significantly affect the optimization process. This paper considers a simple approach to choosing the parameters $CR_{i,g}$, $\varepsilon_{i,g}$, based on jDE [5]:

0) Initialization of parameters $CR_{l,0} = CR_l = 0.9$, $\varepsilon_{q,0} = \varepsilon_{l,0} = \varepsilon_l = 1$.
1) Updating parameters:

$$\varepsilon_{q,g} = \frac{1}{((g-1)NP + q)^{\frac{1}{2D}}}. \qquad (20)$$

$$CR_{l,g} = rand(0,1), \text{ if } rand(0,1) \leq 0.1, \text{ else } CR_l. \qquad (21)$$

$$\varepsilon_{l,g} = \frac{1}{((g-1)NP + l)^{\frac{1}{2}}}, \text{ if } rand(0,1) \leq 0.1, \text{ else } \varepsilon_l, \qquad (22)$$

2) Saving parameters when successfully replacing the parent vector with the test vector:

$$CR_l = CR_{l,g}, \text{ if } J(N_{l,g}) \leq J(\bar{N}_{l,g+1}),\tag{23}$$

$$\varepsilon_l = \varepsilon_{l,g}, \text{ if } J(N_{l,g}) \leq J(\bar{X}_{l,g+1}),\tag{24}$$

where $q = \overline{1, Q_{max}}$, $l = \overline{Q_{max} + 1, NP}$ and $g = \overline{1, G_{max}}$.

The population size is often taken proportional to the dimension of the problem $NP = 10m$ [5].

The parameter ψ_g, which regulates the convergence rate, takes the following value:

$$\psi_g = ((g - 1)NP + 1)^{\frac{1}{\lambda}},\tag{25}$$

where λ is some positive constant.

It was proved that for any $\delta > 0$, the probability of choosing the test vector $\bar{N}_{q,g}$ from the δ-neighborhood of the vector of global maximum N^* tends to unity when g becomes infinitely large [29]. In other words, the probability density of the choice of the test vector $\bar{N}_{q,g}$ tends to the delta-function centered at the vector of the global maximum N^*. Thus, the optimization algorithm converges.

3 Results

To find the optimal weights of the neural network, we used the following evolution methods: SoFDE, original SoFA, the Differential Evolution with different mutation strategies such as DE best 1, DE current to pBest 1, DE current to best 1, DE rand 1 [5,31]. Method parameters and stopping conditions are the same for all algorithms and take the following values: $\lambda = 4$, $m = 153$, $NP = 1530$, $Q_{max} = 153$, $K_{max} = 153 \cdot 10^3$, $p = 15$. We also use the configuration of parameters from jDE [5].

Comparison of the convergence rate for different methods is shown in Fig. 2. Here the average dependence of the found best fitness value on the number of iterations is presented on 15 runs. It can be seen that the SoFDE method provides a higher convergence rate and greater accuracy in finding the maximum fitness.

The optimal pattern of DVM obtained by the SoFDE method is shown in Fig. 3. For comparison, the results of field observations in the Black Sea are also presented here [8]. The highest observed concentration of biomass is marked in gray. It can be seen in Fig. 3 that we get the good approximation of the real observed behavioral strategy of DVM.

We considered the work of the trained neural network with noisy data of the amount of food:

$$F_i(x) = \frac{\tanh(\xi_1(x - c_1)) + 1}{2} + rand(-\Theta_i, \Theta_i),\tag{26}$$

where $i = \overline{1, 2}$ and $\Theta_1 = 0.05$; $\Theta_2 = 0.15$.

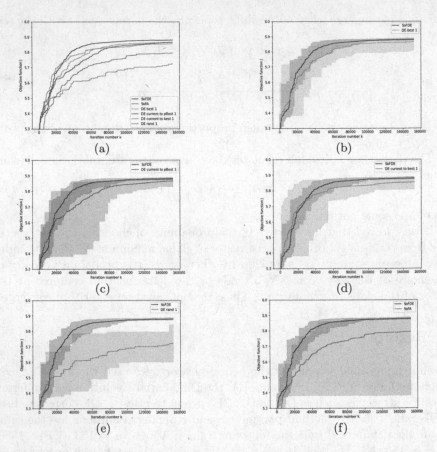

Fig. 2. (a) Comparison of the average convergence rate with its upper and lower boundary for the SoFDE algorithm (blue line) and (b) DE best1; (c) DE current to pBest 1; (d) DE current to best 1; (e) DE rand 1; (f) SoFA. (Color figure online)

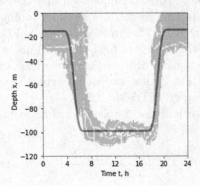

Fig. 3. The echogram and the optimal strategy of DVM obtained by the neural network and value of the fitness function J is 5.894.

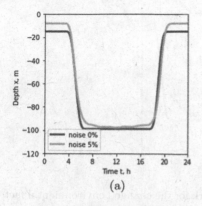 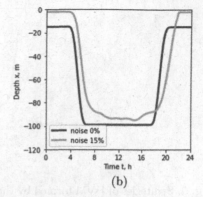

(a)   (b)

Fig. 4. Strategies of DVM formed by the network for the changed environmental factor F with (a) 5% and (b) 15% percent noise. The optimal graph corresponds to the strategy without noise; the simulated graph corresponds to the strategy formed by the neural network.

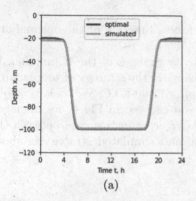 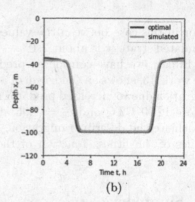

(a)   (b)

Fig. 5. Strategies of DVM formed by the network for the changed environmental factor D with (a) $c_3 = -20$ and (b) $c_3 = -40$. The optimal graph corresponds to the true optimal strategy for these conditions and value of the fitness function J is (a) 5.967, (b) 5.841; the simulated graph corresponds to the strategy formed by the neural network and value of the fitness function J is (a) 5.932, (b) 5.803.

The trajectory formed by the network is shown in Fig. 4. It can be seen that the presence of a small noise changes the trajectory insignificantly; the created feedback permits to form the strategy with noisy values of environmental factors.

Then we have investigated the work of the training network for different environments. We have considered season changes of the D function [27]. We took $c_3 = -20$ and $c_3 = -40$. The obtained patterns of DVM are shown in Fig. 5 by the green line. Here the true optimal strategy is also shown by the blue line. Comparison of the two graphs allows us to evaluate the efficiency of the trained neural network in a changed environment. It can be seen that the network almost exactly forms the trajectory for changed conditions. The relative

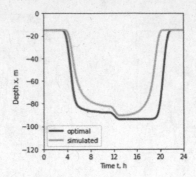

Fig. 6. Strategies of DVM formed by the network for the changed environmental factor P ($c_2 = -90$ before 12:00 UTC and $c_2 = -95$ after). The optimal graph corresponds to the true optimal strategy for these conditions and value of the fitness function J is 6.291; the simulated graph corresponds to the strategy formed by the neural network and value of the fitness function J is 5.894.

error in this case between the values of the fitness function of the optimal and simulated strategy is about 2.1%.

Finally, we have considered predator activity changes of the P function. In the work [13] shows an echogram of a sharp change in the strategy of zooplankton migration due to increased predator activity after 12:00 UTC. We took $c_2 = -90$ before 12:00 UTC and $c_2 = -95$ after. As you can see in Fig. 6, we managed to achieve the described migration. The relative error in this case between the values of the fitness function of the optimal and simulated strategy is about 6.3%.

4   Summary

In this study, we develop a novel approach to model strategies of diel vertical migrations (DVM) of zooplankton. We have constructed feedback that connects the current environmental state and the organism's local behavior. We solved the problem on the base of the neuron network – the multilayer perceptron. Its input is the values of environmental factors at the given point and the given time, its output is the corresponding local displacement of zooplankton. Net weights are selected to provide maximum zooplankton fitness. To train the network, we applied evolution methods of stochastic global optimization: the Differential Evolution and the novel Survival of the Fittest by Differential Evolution algorithm. It was shown that this approach permits to form the optimal behavioral strategy for different environmental conditions, and for noisy values of environmental factors. The obtained trajectory is the good approximation of the real pattern of DVM observed in the Black Sea. It was shown that the Survival of the Fittest by Differential Evolution algorithm provides a higher neural network convergence rate and greater accuracy in finding the maximum fitness.

Our straightforward tests of the new software demonstrated the great potential of the proposed methodology in revealing DVM in the case of noisy and dynamical environmental conditions.

References

1. Archibald, K.M., Siegel, D.A., Doney, S.C.: Modeling the impact of zooplankton diel vertical migration on the carbon export flux of the biological pump. Global Biogeochem. Cycles **33**(2), 181–199 (2019)
2. Arcifa, M.S., Perticarrari, A., Bunioto, T.C., Domingos, A.R., Minto, W.J.: Microcrustaceans and predators: diel migration in a tropical lake and comparison with shallow warm lakes. Limnetica **35**(2), 281–296 (2016)
3. Baioletti, M., Di Bari, G., Milani, A., Poggioni, V.: Differential evolution for neural networks optimization. Mathematics **8**(1) (2020)
4. Birch, J.: Natural selection and the maximization of fitness. Biol. Rev. Camb. Philos. Soc. **91**(3), 712–727 (2015)
5. Brest, J., Maucec, M.S., Bošković, B.: The 100-digit challenge: algorithm jDE100. In: 2019 IEEE Congress on Evolutionary Computation (CEC), pp. 19–26 (2019)
6. Buesseler, K.O., et al.: Revisiting carbon flux through the ocean's twilight zone. Science **316**(5824), 567–570 (2007)
7. Clark, C., Mangel, M.: Dynamic State Variable Models in Ecology. Oxford University Press (2000)
8. Danovaro, R., et al.: Implementing and innovating marine monitoring approaches for assessing marine environmental status. Front. Mar. Sci. **3** (2016)
9. Ducklow, H.W., Steinberg, D.K., Buesseler, K.O.: Upper ocean carbon export and the biological pump. Oceanography **14**(4), 50–58 (2001)
10. Fiksen, O., Giske, J.: Vertical distribution and population dynamics of copepods by dynamic optimization. ICES J. Mar. Sci. **52**, 483–503 (1995)
11. Gabriel, W., Thomas, B.: Vertical migration of zooplankton as an evolutionarily stable strategy. Am. Nat. **132**(2), 199–216 (1988)
12. Gavrilets, S.: Fitness Landscapes and the Origin of Species (MPB-41). Princeton University Press, Princeton (2004)
13. Godø, O.R., et al.: Marine ecosystem acoustics (MEA): quantifying processes in the sea at the spatio-temporal scales on which they occur. ICES J. Mar. Sci. **71**(8), 2357–2369 (2014)
14. Gorban, A.N.: Selection theorem for systems with inheritance. Math. Model. Nat. Phenom. **2**(4), 1–45 (2007)
15. Guerra, D., et al.: Zooplankton diel vertical migration in the Corsica channel (North-Western Mediterranean sea) detected by a moored acoustic doppler current profiler. Ocean Sci. **15**(3), 631–649 (2019)
16. Häfker, N.S., Meyer, B., Last, K.S., Pond, D.W., Hüppe, L., Teschke, M.: Circadian clock involvement in zooplankton diel vertical migration. Curr. Biol. **27**(14), 2194–2201 (2017)
17. Hays, G.C.: A review of the adaptive significance and ecosystem consequences of zooplankton diel vertical migrations. Hydrobiologia **503**(1), 163–170 (2003)
18. Isla, A., Scharek, R., Latasa, M.: Zooplankton diel vertical migration and contribution to deep active carbon flux in the NW Mediterranean. J. Mar. Syst. **143**, 86–97 (2015)

19. Ismailov, V.E.: On the approximation by neural networks with bounded number of neurons in hidden layers. J. Math. Anal. Appl. **417**(2), 963–969 (2014)
20. Kaelo, P., Ali, M.M.: Some variants of the controlled random search algorithm for global optimization. J. Optim. Theory Appl. **130**(2), 253–264 (2006)
21. Kaiser, M., et al.: Marine Ecology: Processes, Systems and Impacts. Oxford University Press, Oxford (2011)
22. Kuzenkov, O., Morozov, A.: Towards the construction of a mathematically rigorous framework for the modelling of evolutionary fitness. Bull. Math. Biol. **81**(11), 4675–4700 (2019)
23. Kuzenkov, O.A., Grishagin, V.A.: Global optimization in Hilbert space. In: AIP Conference Proceedings, vol. 1738, no. 1, p. 400007 (2016)
24. Lampert, W.: The adaptive significance of diel vertical migration of zooplankton. Funct. Ecol. **3**(1), 21–27 (1989)
25. Lee, E.B., Lawrence, M.: Foundations of Optimal Control Theory (1967)
26. Lippmann, R.: An introduction to computing with neural nets. IEEE ASSP Mag. **4**(2), 4–22 (1987)
27. Morozov, A., Kuzenkov, O.A., Arashkevich, E.G.: Modelling optimal behavioural strategies in structured populations using a novel theoretical framework. Sci. Rep. **9**(1), 15020 (2019)
28. Morozov, A.Y., Kuzenkov, O.A.: Towards developing a general framework for modelling vertical migration in zooplankton. J. Theor. Biol. **405**, 17–28 (2016)
29. Morozov, A.Y., Kuzenkov, O.A., Sandhu, S.K.: Global optimisation in Hilbert spaces using the survival of the fittest algorithm. Commun. Nonlinear Sci. Numer. Simul. **103**, 106007 (2021)
30. Parvinen, K., Dieckmann, U., Heino, M.: Function-valued adaptive dynamics and the calculus of variations. J. Math. Biol. **52**(1), 1–26 (2006)
31. Price, K., Storn, R., Lampinen, J.: Differential Evolution: A Practical Approach to Global Optimization (2005)
32. Price, W.L.: A controlled random search procedure for global optimisation. Comput. J. **20**, 367–370 (1977)
33. Price, W.L.: Global optimization by controlled random search. J. Optim. Theory Appl. **40**, 333–348 (1983)
34. Ringelberg, J.: Diel Vertical Migration of Zooplankton in Lakes and Oceans, pp. 1–9. Springer, Dordrecht (2010)
35. Santos, C., Gonçalves, M., Hernández Figueroa, H.E.: Designing novel photonic devices by bio-inspired computing. IEEE Photonics Technol. Lett. **22**, 1177–1179 (2010)
36. Sergeyev, Y.D., Grishagin, V.A.: Parallel asynchronous global search and the nested optimization scheme. J. Comput. Anal. Appl. **3**(2), 123–145 (2001)
37. Sergeyev, Y.D., Strongin, R.G., Lera, D.: Introduction to Global Optimization Exploiting Space-Filling Curves. Springer, New York (2013)
38. Storn, R., Price, K.: Differential evolution - a simple and efficient heuristic for global optimization over continuous spaces. J. Global Optim. **11**(4), 341–359 (1997)
39. Strongin, R.G., Sergeyev, Y.D.: Global Optimization with Non-convex Constraints: Sequential and Parallel Algorithms (2000)
40. He, X., Xu, S.: Feedback process neural networks. In: He, X., Xu, S. (eds.) Process Neural Networks, pp. 128–142. Springer, Cham (2010). https://doi.org/10.1007/978-3-540-73762-9_6
41. Zhigljavsky, A., Žilinskas, A.: Stochastic Global Optimization. Springer, Cham (2008)

Synthesis of Trajectory Planning Algorithms Using Evolutionary Optimization Algorithms

Dmitry Malyshev[1]([✉]) [iD], Vladislav Cherkasov[1] [iD], Larisa Rybak[1] [iD],
and Askhat Diveev[2,3] [iD]

[1] Belgorod State Technological University named after V.G. Shukhov,
Belgorod, Russia
rlbgtu@gmail.com
[2] Federal Research Center "Computer Science and Control"
of Russian Academy of Sciences, Moscow, Russia
[3] RUDN University, Moscow, Russia

Abstract. The article considers the problem of planning the optimal trajectory of a delta robot. The workspace of the robot is limited by the range of permissible values of the angles of the drive revolute joints, interference of links and singularities. Additional constraints related to the presence of obstacles have been introduced. Acceptable values of the robot's input coordinates are obtained based on the inverse kinematics, taking into account the constraints of the workspace, represented as a partially ordered set of integers. For the given initial and final coordinates, a randomly generated family of trajectories belonging to a valid set is obtained. Optimization of each of the trajectories of the family based on evolutionary algorithms is performed. The optimization criterion is a function proportional to the duration of movement along the trajectory. The results of modeling are presented.

Keywords: Trajectory planning · Genetic algorithm · Grey Wolf optimization · Particle swarm optimization

1 Introduction

Robot trajectory planning is essential for avoiding various obstacles and obtaining the optimal path in terms of various criteria. Currently, there are a number of methods that can be used for trajectory planning. Some of the well-known methods are based on route networks, including the method based on the application of Visibility Graphs [1]. An alternative method for determining routes is based on the use of Generalized Voronoi Diagrams [2]. Thus, the method of uninformed search allows implementing breadth-first search, depth-first search, search by cost criterion [3]. Heuristic pathfinding algorithms are designed to quickly find a

This work was supported by the state assignment of Ministry of Science and Higher Education of the Russian Federation under Grant FZWN-2020-0017.

route in a graph by propagating towards more promising vertices [4,5]. In recent years, a number of methods for trajectory planning have been proposed. A two-stage method for planning the trajectory of two mobile manipulators for joint transportation in the presence of static obstacles is considered in [6]. At the first stage of path planning in progress, the shortest possible path between the initial and target configuration is executed in the workspace. The second stage consists in calculating a sequence of time-optimal trajectories for passing between consecutive points of the path, taking into account non-holonomic constraints and maximum permissible joint accelerations. A new method of spatial decomposition is presented in [7]. It was applied to define a space for trajectory planning, called RV-space. In [8], the method of directed reachable volumes (DRVs) is presented, which makes it possible to obtain the area taking into account constraints on the positions of the robot's links and end-effector. The current work considers the application of evolutionary and bio-inspired algorithms for planning the trajectory of a delta robot, taking into account the limitations of its workspace.

2   Setting an Optimization Problem

The delta robot [9] with 3 degrees of freedom is showed on Fig. 1. The end-effector of delta robot is the center P of the moving platform with x_P, y_P, z_P coordinates.

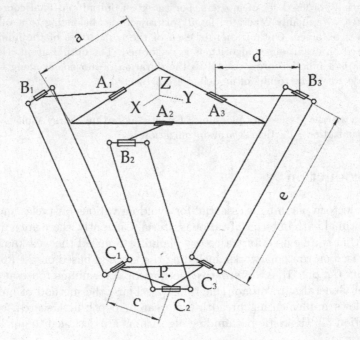

Fig. 1. Structure of the delta robot

The input coordinates for a delta robot are the angles θ_i of rotation of the drive revolute joints A_i. Inverse kinematics allows to transfer the described constraints of the workspace to the space of input coordinates. Inverse kinematics [10]:

$$\theta_i = 2\tan^{-1}\left(\frac{-F_i \pm \sqrt{E_i^2 + F_i^2 + G_i^2}}{G_i - E_i}\right) \tag{1}$$

where E_i, F_i and G_i defined as $E_1 = 2d\left(y_P + \frac{a-2c}{2\sqrt{3}}\right), F_1 = 2z_P d$,

$$G_1 = v + \left(\frac{a-2c}{2\sqrt{3}}\right)^2 + 2y_P\left(\frac{a-2c}{2\sqrt{3}}\right) - e^2, E_2 = -d\left(\sqrt{3}\left(x_P + \frac{2c-a}{4}\right) + y_P + \frac{2c-a}{4\sqrt{3}}\right),$$

$$F_2 = 2z_P d, G_2 = v + \left(\frac{2c-a}{4}\right)^2 + \left(\frac{2c-a}{4\sqrt{3}}\right)^2 + 2\left(x_P\left(\frac{2c-a}{4}\right) + y_P\left(\frac{2c-a}{4\sqrt{3}}\right)\right) - e^2,$$

$$F_3 = 2z_P d, G_3 = v + \left(\frac{2c-a}{4}\right)^2 + \left(\frac{2c-a}{4\sqrt{3}}\right)^2 + 2\left(y_P\left(\frac{2c-a}{4\sqrt{3}}\right) - x_P\left(\frac{2c-a}{4}\right)\right) - e^2, E_3$$

$$= d\left(\sqrt{3}\left(x_P - \frac{2c-a}{4}\right) - y_P - \frac{2c-a}{4\sqrt{3}}\right), v = x_P^2 + y_P^2 + z_P^2 + d^2$$

An arbitrary trajectory can be represented as a set of movements (steps), during which the revolute joint drives operate at a constant angular velocity and the movement in the space of input coordinates is rectilinear. In order to reduce the duration of such small movements, the highest of the angular speeds of the drives at each step should correspond to the maximum possible. The duration of a move is proportional to the sum of individual steps defined in accordance with the Chebyshev metric:

$$t = \frac{1}{\omega_{\max}} \sum_{i=1}^{n} \rho_i \tag{2}$$

where $\rho_i = \max_{j \in \{1,2,...,m\}} |\theta_{i,j} - \theta_{i-1,j}|$ - according to Chebyshev distance between the points of the beginning of $C_{i-1}(\theta_{i-1,1}, \theta_{i-1,2}, .., \theta_{i-1,m})$ and end with $C_{i-1}(\theta_{i,1}, \theta_{i,2}, .., \theta_{i,m})$ i-th step; m is the number of the input coordinates (for the Delta robot m = 3); ω_{max} is the maximum angular velocity of the drive revolute joints. However, the direct application of this indicator as a criterion function for optimizing the trajectory is impractical, since the Chebyshev metric introduces significant ambiguity. On the other hand, using the criterial "usual" length of the trajectory (the sum of the Euclidean lengths of all steps) as a criterion function also not allowed. The duration of movement in this case may be far from optimal as a result. Therefore, it is proposed to supplement the criterion function with a Euclidean metric taken with a small weighting coefficient ϵ:

$$F = \sum_{i=1}^{n} \left(\max_{j \in \{1,2,...,m\}} |\theta_{i,j} - \theta_{i-1,j}| + \varepsilon \sqrt{\sum_{j=1}^{m} (\theta_{i,j} - \theta_{i-1,j})^2} \right) \rightarrow \min \tag{3}$$

The approach based on the application of a criterion function of this type was successfully tested by the authors to optimize the 3-RPR mechanism's trajectory

[11]. Optimization should be carried out with constraints on the size of the workspace. In the framework of previous works, the authors proposed to use the representation of the workspace in the form of a partially ordered set of integers A_P [12]. Therefore, checking the optimization constraint consists of two steps.

The First Stage. Definition of the Set B of Trajectory Coordinates in the Space of Integers

For this purpose, an algorithm based on a modification of the algorithm is developed Bresenham's algorithm [13], which assumes that the trajectory is represented as a polyline consisting of many segments. In [14], a modification of the algorithm was proposed for the 3D case, but the coordinates of the beginning and end of the segments belong to the space of integers, which leads to a displacement of the trajectory segment and the set B (Fig. 2). Cells that intersect the orthosis are highlighted in red for coordinates represented as integers, yellow for coordinates represented as real numbers, and orange for both cases. As you can see from the figure, using integer coordinates does not allow you to accurately determine the set B.

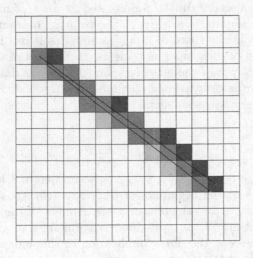

Fig. 2. Offset of the trajectory segment depending on the source data

Modifying the algorithm considering the original data that belongs to the 3D space of real numbers (coordinates of the beginning x_1, y_1, z_1 and end x_2, y_2, z_2 segments). In this case, the coordinates must correspond to the covering set of the workspace, represented as a partially ordered set of integers, respectively, they must be obtained taking into account the accuracy of the approximation Δ_j and the displacement k_j along the j coordinate axes by the formula

$$x_i = \frac{\theta_{i,1} + k_1}{\Delta_1}, \quad y_i = \frac{\theta_{i,2} + k_2}{\Delta_2}, \quad z_i = \frac{\theta_{i,3} + k_2}{\Delta_2} \tag{4}$$

The algorithm works as follows:

Input: $x_1, y_1, z_1, x_2, y_2, z_2$

1: $\delta_x = x_2 - x_1, \delta_y = y_2 - y_1, \delta_z = z_2 - z_1$

• 2: **if** $\delta_x = 0$ **then** $x_1 = [x_1], x_2 = [x_2]$ **end if**

3: **if** $\delta_y = 0$ **then** $y_1 = [y_1], y_2 = [y_2]$ **end if**

4: **if** $\delta_z = 0$ **then** $z_1 = [z_1], z_2 = [z_2]$ **end if**

5: $B = B \cup \{[x_1], [y_1], [z_1]\}$

6: **if** $\delta_x < 0$ **then** $x = \lfloor x_1 + 0.5 \rfloor - 0.5$ **else** $x = \lceil x_1 - 0.5 \rceil + 0.5$ **end if**

7: **if** $\delta_y < 0$ **then** $y = \lfloor y_1 + 0.5 \rfloor - 0.5$ **else** $y = \lceil y_1 - 0.5 \rceil + 0.5$ **end if**

8: **if** $\delta_z < 0$ **then** $z = \lfloor z_1 + 0.5 \rfloor - 0.5$ **else** $z = \lceil z_1 - 0.5 \rceil + 0.5$ **end if**

9: **while** $x \cdot \text{sign}(\delta_x) \le x_2 \cdot \text{sign}(\delta_x)$ **do**

10:      $x_Z = x + 0.5 \cdot \text{sign}(\delta_x), y_R = y_1 + \delta_y \cdot (x - x_1)/\delta_x, z_R = z_1 + \delta_z \cdot (x - x_1)/\delta_z$

11:      **if** $(\lfloor y_R \rfloor \ne y_R - 0.5)$ **or** $(\text{sign}(y_R) = \text{sign}(\delta_y))$ **then**

12:          $y_Z = [y_R]$

13:      **else**

14:          $y_Z = \text{sign}(y_R) \cdot \lfloor |y_R| \rfloor$

15:      **end if**

16:      **if** $(\lfloor z_R \rfloor \ne z_R - 0.5)$ **or** $(\text{sign}(z_R) = \text{sign}(\delta_z))$ **then**

17:          $z_Z = [z_R]$

18:      **else**

19:          $z_Z = \text{sign}(z_R) \cdot \lfloor |z_R| \rfloor$

20:      **end if**

21:      $B = B \cup \{x_Z, y_Z, z_Z\}, x = x + \text{sign}(\delta_x)$

22: **end while**

23: **while** $y \cdot \text{sign}(\delta_y) \le y_2 \cdot \text{sign}(\delta_y)$ **do**

24:      $x_R = x_1 + \delta_x \cdot (y - y_1)/\delta_y$

25:      **if** $\lfloor x_R \rfloor \ne x_R - 0.5$ **then**

26:          $x_Z = [x_R], y_Z = y + 0.5 \cdot \text{sign}(\delta_y), z_R = z_1 + \delta_z \cdot (x - x_1)/\delta_z$

27:          **if** $(\lfloor z_R \rfloor \ne z_R - 0.5)$ **or** $(\text{sign}(z_R) = \text{sign}(\delta_z))$ **then**

28:              $z_Z = [z_R]$

29:          **else**

30:              $z_Z = \text{sign}(z_R) \cdot \lfloor |z_R| \rfloor$

31:          **end if**

32:          $B = B \cup \{x_Z, y_Z, z_Z\}$

33:      **end if**

34:      $y = y + \text{sign}(\delta_y)$

35: **end while**

36: **while** $z \cdot \text{sign}(\delta_z) \le z_2 \cdot \text{sign}(\delta_z)$ **do**

37:      $x_R = x_1 + \delta_x \cdot (z - z_1)/\delta_z, y_R = y_1 + \delta_y \cdot (z - z_1)/\delta_z,$

38:      **if** $\lfloor x_R \rfloor \ne x_R - 0.5$ **and** $\lfloor y_R \rfloor \ne y_R - 0.5$ **then**

39:          $x_Z = [x_R], y_Z = [y_R], z_Z = z + 0.5 \cdot \text{sign}(\delta_z)$

40:          $B = B \cup \{x_Z, y_Z, z_Z\}$

41:      **end if**

42:      $z = z + \text{sign}(\delta_z)$

43: **end while**

The algorithm for determining the coordinates of the trajectory in the space of integers for a 2D trajectory assumes the exclusion of the z dimension. Otherwise, it is similar to the algorithm given above for the 3D case.

The Second Stage. Checking Whether the Resulting set B Belongs to the Workspace Set A_θ

Thus the optimization constraint condition has the form

$$B_i \subset A_\theta, i \in 1, .., n \tag{5}$$

where n is the number of segments that make up the trajectory.

Thus, the optimization problem looks like this.

- parameters: coordinates of intermediate points of the trajectory $x_i, y_i, z_i, i \in 1, .., (n-1)$. For a delta robot, the coordinates are the rotation angles of the drive revolute joints, i.e. $[(x_i y_i z_i)]^T = [(\theta_{i,1} \theta_{i,2} \theta_{i,3})]^T$.
- parameter change range: overall dimensions of the workspace in the space of input coordinates $\theta_{i,j} \in [\theta_{j,min}; \theta_{j,max}]$.

It should be noted that the optimization parameters change in the space of real numbers. The transition to the space of integers to calculate the cells checked at the second stage is carried out using the formula (4) and the modified Bresenham's algorithm.

- criterion: the function F calculated by formula (3).
- constraint: condition (5).

To increase the efficiency of optimization in the presence of obstacles, we transfer the optimization constraint to the criterion function

$$F' = F + \sum_{i=1}^{n} \left(\vartheta_i \left(p_1 \sqrt{\sum_{j=1}^{m} (\theta_{i,j} - \theta_{i-1,j})^2} + p_2 \right) \right) \rightarrow \min \tag{6}$$

where p_1, p_2 are the penalty coefficient, and ϑ_i is the Heaviside function:

$$\vartheta_i = \begin{cases} 0, \text{if } B_i \subset A \\ 1 - \text{otherwise} \end{cases} \tag{7}$$

3   Algorithms for a Path Optimization

The choice of algorithms is justified by their efficiency and high level of applicability to a number of different problems. However, the authors do not conclude that these algorithms are better than other evolutionary algorithms for solving this particular problem. The purpose of this investigation is an initial assessment of the applicability of some of the most widely used evolutionary and bio-inspired algorithms for optimizing a trajectory within a workspace represented as a partially ordered set of numbers. This creates the prerequisites for further in-depth research, including a comparative analysis of the application of a larger number of algorithms for this problem and the selection of their parameters.

We apply the following evolutionary and bio-inspired algorithms to solve optimization problem.

3.1   Genetic Algorithm (GA)

The basic principles of GA were first rigorously formulated by Holland [15]. The GA works with a population of "individuals", each representing a possible solution to a given problem. Genetic algorithms are widely applied, including for the synthesis of a control system for robots [16], for planning the trajectory of collaborative robots [17]. We use a modification of the genetic algorithm described earlier in [18]. To speed up the algorithm, we apply parallel computing (Fig. 3). Dashed lines indicate areas where calculations are performed simultaneously.

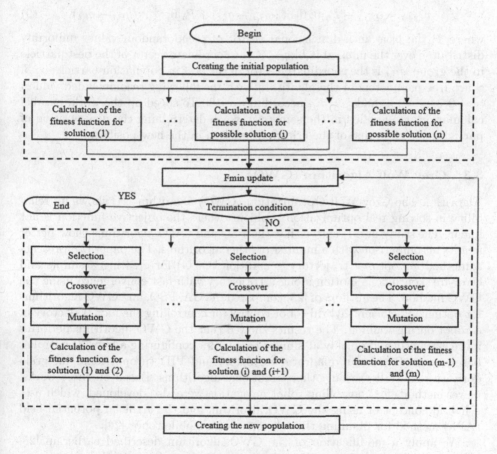

Fig. 3. A genetic algorithm using parallel computing

3.2   Particle Swarm Algorithm (PSO)

Particle Swarm Optimization (PSO) is a widely applied two-component swarm-based evolutionary optimization method [19, 20]. The particle swarm algorithm solves the problem by having a population of potential solutions, here called particles, and moving these particles in the search space according to a simple mathematical formula over the position and velocity of the particle. The movement of each particle depends on its local best-known position, but is also directed to the best-known positions in the search space, which are updated as other particles find better positions. This is expected to move the swarm towards better solutions [18]. New position of the s_i particle At time t is determined by the vector of its coordinates q_i, and its velocity by the vector ϑ_i:

$$q_{i,t+1} = q_{i,t} + v_{i,t+1}; \tag{8}$$

$$v_{i,t+1} = \alpha v_{i,t} + P_n[0; \beta] \times (q_{G,t} - q_{i,t}) + P_n[0; \gamma] \times (q_{P,t} - q_{i,t}) \tag{9}$$

where P_n [a; b] is an n-dimensional vector of pseudorandom values uniformly distributed over the interval [a,b];$q_{G,t}$ is the coordinate vector of the best particle in the group; $q_{B,t}$ is the coordinate vector of the best in population particle; α, β, γ are free parameters of the algorithm with the following recommended values: $\alpha = 0.7$, $\beta = 1.4$, and $\gamma = 1.4$ according to [18]. To speed up the work, parallel calculations are applied in the same way when determining the new position of particles and the value of the criterion function in the new position.

3.3   Grey Wolf Algorithm (GWO)

Algorithm The Grey Wolf Optimization (GWO) algorithm [21] shows its reliability in solving real optimization problems where the objective function is not linear. The study in [21] shows that the PSO and GWO algorithms show better results in comparison with a number of other algorithms. The paper [22] presents a method for optimal trajectory generation (OTG) for creating a simple and error-free continuous motion along a trajectory with fast convergence using the GWO method. The authors of [23] compared the GA, PSO, and GWO algorithms for optimizing efficient hybrid robot control for controlling the foot trajectory of a robot during walking. The results showed that the GWO algorithm performs more efficiently and quickly at similar torques for configuring a hybrid controller based on LQR (Linear quadratic regulator) and PID (proportional–integral–derivative controller) than other traditional algorithms. Based on these works, newer methods for controlling robot navigation were also developed, which uses a hybrid concept of using the GWO algorithm and the artificial potential field (APF) method for planning the trajectory of a mobile robot [24].

We apply a modification of the GWO algorithm described earlier in [25] using parallel calculations to modify the parameters of a possible solution and determine the value of the fitness function.

4   Numerical Results

The problem of determining the workspace A_P for a delta robot in the space of output coordinates is considered by the authors in [26]. The constraints of the workspace A_P are transferred from the space of output coordinates x_P, y_P, z_P

to the space of input coordinates θ_i using the solution of the inverse kinematics (1). The workspace A_θ in the input coordinate space is similarly represented as a partially ordered set of integers after the transfer. An object was added as an obstacle with a parallelepiped shape (Fig. 4a). Elements of the covering set of the workspace in the output coordinate space (Fig. 4b) belonging to the obstacle were excluded $A'_P = A_P \backslash C$. The transfer of constraints $A'_P \rightarrow A_\theta$ (Fig. 4c) was performed for the updated set A'_P.

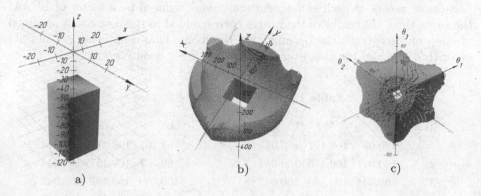

Fig. 4. Additional boundaries, related to the overall dimensions of the obstacle: a) obstacle C; b) Workspace set A_P; c) A'_P; c) A_θ

We perform trajectory optimization using the above-mentioned algorithms for the 2D and 3D case of forming a trajectory inside the workspace of a delta robot, as well as for a randomly generated 2D contour with a large number of obstacles. A C++ software package has been developed for this purpose. Parallel computing is implemented using the OpenMP library. Visualization of 2D results is performed using developed Python scripts Python (Matplotlib and JSON libraries). Visualization of 3D results is performed by exporting an ordered set of integers describing the workspace in STL format and arrays of co-ordinates of trajectory points in JSON format, and then importing data in the Blender software package using developed Python scripts.

4.1 2D Case

Let's make a slice of the workspace A_θ of the delta robot in the space of input coordinates θ_1 θ_2 θ_3, taking the angle $\theta_1 = 0$. In this case, the set A_θ will be 2D (Fig. 5). We assume that the starting point of the trajectory is $\theta_{1,2} = -70°$, $\theta_{1,3} = 20°$, the end point is $\theta_{n,2} = 50°$, $\theta_{n,3} = 0°$, and the number of vertices of the trajectory is n = 3. Accordingly, the number of optimization parameters $p = 2n = 6$. The weight coefficient $\epsilon = 0,1$, the penalty coefficients: $p_1 = 5$, $p_2 = 500$. Parameters of the GA algorithm: the number of individuals in the initial population H = 1000, the number of generations W = 250, the number of crossovers at each iteration $S_{GA} = 500$, the number of possible values of each of the parameters $g = 2^{25}$, the probability of mutation $p_m = 70\%$. Parameters of the GWO algorithm: H = 1000, W = 250, number of new individuals at each iteration $S_{GWO} = 1000$. Parameters of the PSO algorithm: H = 1000, W = 250, number of groups G = 2, values of free parameters $\alpha = 0.7$, $\beta = 1.4$, $\gamma = 1.4$. Optimization for each of the tests was performed in four stages. At the first stage, the range of parameters was changed to the ranges corresponding to the overall dimensions of the workspace for each of the coordinates. The parameter ranges at each subsequent stage were reduced by a factor of 10^2. At the same time, the center of the ranges corresponded to the best result obtained at the previous stage. The optimization results are shown in Table 1. The PSO algorithm provides the best average value of the criterion function.

Table 1. Results table for the 2D case

Trials	GA	GWO	PSO	Trials	GA	GWO	PSO
1	160,089	160,197	160,044	6	160,107	160,378	160,044
2	160,821	160,145	160,044	7	160,007	160,457	160,044
3	160,344	160,340	160,045	8	160,060	160,397	160,042
4	160,048	160,228	160,040	9	160,198	160,212	160,045
5	160,123	160,070	160,044	10	160,471	160,189	160,044
				Avg. values	160,227	160,261	160,044

The obtained trajectories for all tests are almost identical. Examples of trajectories for the first and second tests are shown in Fig. 5. In the second test, the GWO algorithm obtained the trajectory with the largest value of the criterion function. This is clearly seen in Fig. 5b.

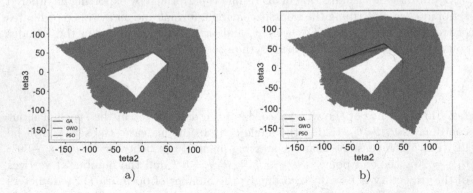

Fig. 5. Results of optimization of the test path: a) 1, b) 2.

The convergence graphs for each of the algorithms are shown in Fig. 6. The minimum value of the criterion function obtained as a result of optimization is applied as the reference value of the function. The minimum value of the criterion function in one of the tests was obtained using the GA algorithm, but in other tests, the PSO algorithm has better convergence rates.

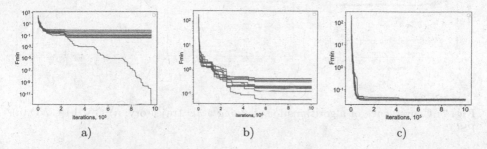

a)   b)   c)

Fig. 6. Convergence of algorithms for planning a 2D trajectory: a) GA, b) GWO, c) PSO

4.2 3D Case

A computational experiment is performed, which consists in planning the trajectory inside the 3D workspace B_θ of a delta robot in the space of input coordinates, taking into account the obstacle, shown in Fig. 4c. Set the starting and ending points of the trajectory in the output coordinate space: $x_{p1} = 250$ mm, $y_{p1} = 250$ mm, $z_{p1} = -500$ mm, $x_{p2} = -270$ mm, $y_{p2} = -270$ mm, $z_{p2} = -450$ mm, and the number of vertices of the trajectory $n = 3$. Accordingly, the number of optimization parameters p $= 3n = 9$. Let's take the parameters of the algorithms $H = 250, W = 200, S_{GA} = 125$, and $S_{GWO} = 250$. The remaining parameters of the computational experiment coincide with the 2D case. Optimization for each of the tests is performed in four stages, similar to the 2D case. The optimization results are shown in Table 2. In this case, The GA algorithm showed the best average value of the criterion function.

Table 2. Results table for the 3D case

Trials	GA	GWO	PSO	Trials	GA	GWO	PSO
1	152,499	149,852	149,737	6	131,130	149,863	149,891
2	151,070	150,136	130,477	7	130,719	130,433	130,459
3	131,368	130,477	149,778	8	150,506	131,111	149,915
4	137,201	150,394	130,385	9	152,649	149,941	149,914
5	149,876	150,427	149,674	10	149,772	150,083	149,769
				Avg. values	143,679	144,272	144,000

The convergence graphs are shown in Fig. 7. The minimum value of the criterion function is obtained using the PSO algorithm.

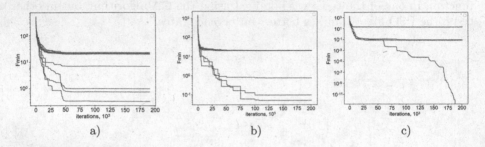

a)   b)   c)

Fig. 7. Convergence of algorithms for planning a 3D trajectory: a) GA, b) GWO, c) PSO.

Figure 8 shows the trajectories for Test 3 inside the workspace. As can be seen from the figure, the PSO algorithm found only a local minimum of the criterion function for avoiding the obstacle.

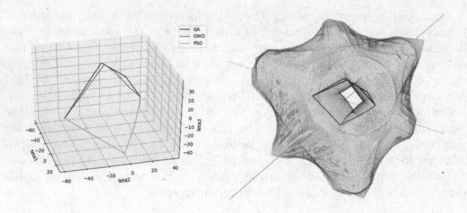

Fig. 8. Results of trajectory optimization

4.3 Trajectory Planning When There are a Large Number of Obstacles

In the first two cases, the trajectory was planned inside the workspace of the delta robot in the presence of a single obstacle. To test the algorithms on the problem of planning a trajectory with a large number of obstacles, a 2D domain was generated, similarly represented as an ordered set of integers. During the experiment, 10 tests were performed similarly for the following initial data: $x_1 = -95$ mm,$y_1 = -95$ mm, $x_2 = 85$ mm, $y_2 = 85$ mm, the number of vertices of the trajectory $n = 7$. Accordingly, the number of optimization parameters $p = 2n = 14$. We assume the parameters of the algorithms $H = 2000, W = 1000, S_{GA} = 1000, S_{GWO} = 2000, p_m = 90\%$. The other parameters were not changed. Optimization for each of the tests was performed in two stages, rather than four. Figure 9 shows examples of the trajectories obtained as a result of optimization. In Fig. 9a, the path that allows you to avoid all obstacles is obtained only for the GA algorithm, in Fig. 9b and c - by the GA and GWO algorithms, in Fig. 9d - by all algorithms. Figure 10 shows an example of the convergence graph of the algorithms. As a result of performing 10 tests for each of the algorithms, the GA algorithm showed the best results GA (Fig. 10a), each time reaching a trajectory that allows to avoid all obstacles. The GWO algorithm (Fig. 10b) it allowed to exclude the interference with an obstacle in 4 cases, and the PSO algorithm-only in one case.

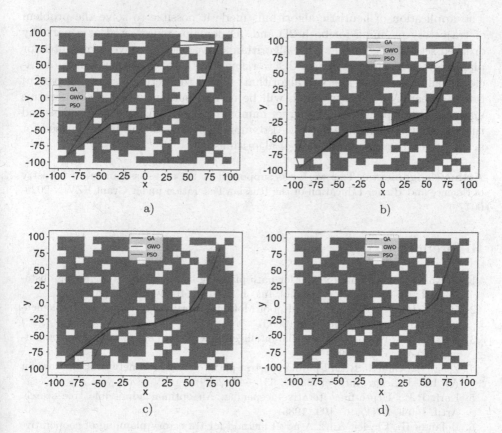

Fig. 9. The result of planning a trajectory in the presence of a large number of obstacles

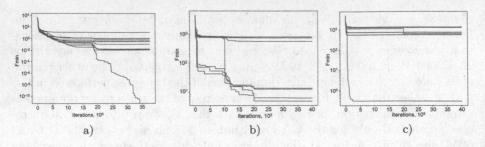

Fig. 10. Convergence of algorithms when planning a trajectory with a large number of obstacles: a) GA, b) GWO, c) PSO.

5   Conclusion

The application of heuristic algorithms made it possible to solve the problem of trajectory planning for both 2D and 3D domains represented as a partially ordered set of integers. The PSO algorithm showed better convergence rates for planning a trajectory within a 2D workspace of a robot with a single obstacle. In all other cases, the GA algorithm showed the better results. As part of future research in-depth research will be carried out, including a comparative analysis of the application of a larger number of algorithms for this problem and the selection of their parameters. Also, more experiments will be performed for accurate comparative evaluation of algorithms.

Acknowledgements. This work was supported by the state assignment of Ministry of Science and Higher Education of the Russian Federation under Grant FZWN-2020-0017.

References

1. El Khaili, M.: Visibility graph for path planning in the presence of moving obstacles. IRACST - Eng. Sci. Technol. Int. J. (ESTIJ) **4**(4), 118–123 (2014)
2. Choset, H., et al.: Principles of Robot Motion-Theory, Algorithms, and Implementation. MIT Press, Cambridge (2005)
3. Russell, S.J., Norvig, P.: Artificial intelligence: a modern approach. Neurocomputing **9**(2), 215–218 (1995)
4. Zeng, W., Church, R.L.: Finding shortest paths on real road networks: the case for A*. Int. J. Geogr. Inf. Sci. **23**(4), 531–543 (2009)
5. Korf, R.E.: Depth-first iterative-deepening. An optimal admissible tree search. Artif. Intell. **27**(1), 97–109 (1985)
6. Bolandi, H., Ehyaei, A.F.: A novel method for trajectory planning of cooperative mobile manipulators. J. Med. Signals Sens. **1**(1), 24–35 (2011)
7. Völz, A., Graichen, K.: An optimization-based approach to dual-arm motion planning with closed kinematics. In: Proceedings of the 2018 IEEE/RSJ International Conference on Intelligent Robots and Systems, IROS, pp. 8346–8351. IEEE (2018)

8. McMahon, T., Sandstrom, R., Thomas, S., Amato, N.M.: Manipulation planning with directed reachable volumes. In: IEEE International Conference on Intelligent Robots and Systems 2017, pp. 4026–4033 (2017)
9. Clavel, R.: Conception d'un robot parallèle rapide à 4 degrés de liberté. Ph.D. Thesis, EPFL, Lausanne, Switzerland (1991)
10. Williams II, R.L.: The delta parallel robot: kinematics solutions. www.ohio.edu/people/williar4/html/pdf/DeltaKin.pdf. Accessed 9 Oct 2022
11. Khalapyan, S., Rybak, L., Malyshev, D., Kuzmina, V.: Synthesis of parallel robots optimal motion trajectory planning algorithms. In: IX International Conference on Optimization and Applications (OPTIMA 2018), pp. 311–324 (2018)
12. Rybak, L., Malyshev, D., Gaponenko, E.: Optimization algorithm for approximating the solutions set of nonlinear inequalities systems in the problem of determining the robot workspace. Commun. Comput. Inf. Sci. **1340**, 27–37 (2020)
13. Rogers, D.: Procedural Elements for Computer Graphics. McGraw-Hill (1985)
14. line3D - 3D Bresenham's (a 3D line drawing algorithm). https://ftp.isc.org/pub/usenet/comp.sources.unix/volume26/line3d. Accessed 9 Oct 2022
15. Holland, J.H.: Adaptation in Natural and Artificial Systems. MIT Press, Cambridge (1975)
16. Diveev, A.: Cartesian genetic programming for synthesis of control system for group of robots. In: 28th Mediterranean Conference on Control and Automation, MED 2020, pp. 972–977 (2020)
17. Zanchettin, A.M., Messeri, C., Cristantielli, D., Rocco, P.: Trajectory optimisation in collaborative robotics based on simulations and genetic algorithms. Int. J. Intell. Robot. Appl. **9**, 707–723 (2022)
18. Diveev, A.I., Konstantinov, S.V.: Evolutionary algorithms for the problem of optimal control. RUDN J. Eng. Res. **18**(2), 254–265 (2017)
19. Kennedy, J.; Eberhart, R.: Particle swarm optimization. In: Proceedings of IEEE International Conference on Neural Networks, vol. IV, pp. 1942–1948 (1995)
20. Shi, Y.; Eberhart, R.C.: A modified particle swarm optimizer. In: Proceedings of IEEE International Conference on Evolutionary Computation, pp. 69–73 (1998)
21. Mirjalili, S., Mirjalili, S.M., Lewis, A.: Grey wolf optimizer. Adv. Eng. Softw. **69**, 46–61 (2014)
22. Choubey, C., Ohri, J.: Optimal trajectory generation for a 6-DOF parallel manipulator using grey wolf optimization algorithm. Robotica **39**(3), 411–427 (2021)
23. Sen, M.A., Kalyoncu, M.: Grey wolf optimizer based tuning of a hybrid LQR-PID controller for foot trajectory control of a quadruped robot. Gazi Univ. J. Sci. **32**(2), 674–684 (2019)
24. Zafar, M.N., Mohanta, J.C., Keshari, A.: GWO-potential field method for mobile robot path planning and navigation control. Arab. J. Sci. Eng. **46**(8), 8087–8104 (2021). https://doi.org/10.1007/s13369-021-05487-w
25. Diveev, A.I., Konstantinov, S.V.: Optimal control problem and its solution by grey wolf optimizer algorithm. RUDN J. Eng. Res. **19**(1), 67–79 (2018)
26. Malyshev, D., Rybak, L., Carbone, G., Semenenko, T., Nozdracheva, A.: Optimal design of a parallel manipulator for aliquoting of biomaterials considering workspace and singularity zones. Appl. Sci. **12**(4), 2070 (2022)

Application of Attention Technique
for Digital Pre-distortion

Aleksandr Maslovskiy[1]([✉])[iD], Aleksandr Kunitsyn[4][iD],
and Alexander Gasnikov[1,2,3][iD]

[1] Moscow Institute of Physics and Technology, Dolgoprudny, Russia
aleksandr.maslovskiy@phystech.edu
[2] Institute for Information Transmission Problems RAS, Moscow, Russia
[3] Caucasus Mathematical Center, Adyghe State University, Maykop, Russia
[4] Moscow State University, Moscow, Russia

Abstract. This paper reviews application of modern optimization methods for functionals describing digital predistortion (DPD) of signals with orthogonal frequency division multiplexing (OFDM) modulation. The considered family of model functionals is determined by the class of cascade Wiener–Hammerstein models, which can be represented as a computational graph consisting of various nonlinear blocks. To assess optimization methods with the best convergence depth and rate as a properties of this models family we multilaterally consider modern techniques used in optimizing neural networks and numerous numerical methods used to optimize non-convex multimodal functions.

The research emphasizes the most effective of the considered techniques and describes several useful observations about the model properties and optimization methods behavior.

Keywords: Digital pre-distortion · TMPA · IGIRNN

1 Introduction

For this moment there are a lot of base station, which are really necessary to realize really good wireless connection. For now there is a need to create more high speed signal, which will allows us to download huge size data really fast. As a result, modern signals should have really complex modulation and high frequency bandwidth(for example third generation partnership project (3GPP) set the frequency limits in the 5G new radio (NR) standard, namely, frequency range 1 (FR1: 0.4–7.1 GHz range) and frequency range 2 (FR2: 24–52.6 GHz) [8]). It is widely assumed that wireless communication in the fifth generation would have greatly enhanced capacity and communication rate. The envelope amplitude distortion is caused by the amplitude variations on the input while amplifying the departure from a straight line input-output transfer function in the cut-off and saturation regions [2,6] (Fig. 1).

N. Olenev et al. (Eds.): OPTIMA 2022, CCIS 1739, pp. 168–182, 2022.
https://doi.org/10.1007/978-3-031-22990-9_12

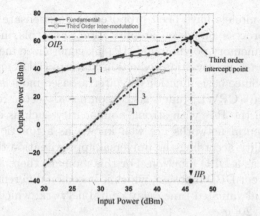

Fig. 1. AM-AM amplitude characteristic of input/output signal

There are many approaches aiming to linearize the RF PA and keep a high efficiency at the same time, including feedback linearization [12], feedforward linearization [13], analog predistortion and digital predistortion (DPD) [7]. DPD is often regarded as the most powerful linearization technology because to its versatility and great performance. The correct modeling and linearization of wideband RF PA is now of great interest to academics and engineers. In this case we used next approach of using DPD technique (Fig. 2):

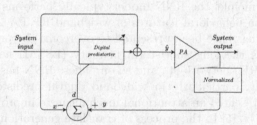

Fig. 2. Used technique of digital-pre-distortion

From the mathematical point of view this approach can be following formulation:

$$\frac{1}{2}\|PA(DPD(x) + x) - x\|_2^2 \to \min_{DPD(x)}$$

All distortions of signal could be presented as additive changes. So, optimization problem could be reformulated in the following form

$$\frac{1}{2}\|DPD_\theta(x) - d\|_2^2 \to \min_{\theta \in \Theta}.$$

where $d = y - x$.

A lot of DPD models have been proposed to compensate the Imd's which is generated in process of power amplification. For this moment, Volterra-based models, like memory polynomial (MP) [5], generalized memory polynomial (GMP) [17] and dynamic deviation reduction Volterra (DDR) model [24] are the most widely used. Since basis function of Volterra-based models, based on canonical piecewise-linear (CPWL) functions, decomposed vector rotation (DVR) is proposed to model the PAs with strong nonlinearity such as envelope tracking (ET) PAs [23]. Neural networks are well known as a function with nonlinear approximation ability according the universal approximation theorem [11], so it can be considered as a DPD function. For this moment there are a lot of implementation of different DPD function, which is based on different neural networks approaches, like real-valuated time delay neural network, which is based on multilayer perceptron [20] etc.

However, in wideband communication circumstances, the memory effect and nonlinearities of RF PAs are much more severe and complicated, resulting in MLP architecture's linearization performance being significantly worse. It should be observed that when bandwidth increases, the memory effect of power amplifiers increases dramatically [21].

As commanly known, because of its capacity to replicate memory effect, the recurrent neural network (RNN) can be used for prediction of time series data [22], and as a result it has been employed to model the behavior of the PA [16]. The RNN, on the other hand, has a difficulty with vanishing gradients, meaning that the model cannot adequately reflect the long-term memory effect [1,9]. As a result, while the RNN model has a higher modeling capacity than the MLP model in principle, the RNN model typically performs worse in practice when modeling the behavioral features of wideband RF PAs. A number RNN-based variant models have been presented to overcome the problem of gradient vanishing, such as the long short-term memory (LSTM) network [10], gated recurrent unit (GRU) network [4], and so on. These RNN-based variant models perform admirably, particularly in wideband digital predistortion [21]–[3]. In 2014 there were presented an attentional mechanism to improve neural machine translation (NMT). [15] In the process of translate generation decoder of model selectively focused on the parts of the source. Because of RNN can be used in time series processing, attention mechanism there were implemented added as well [18]. Currently, there were implemented attention approach for DPD model [21]. In this case researchers implemented original approach of attention mechanism based on latent vectors of context information from LSTM model. But this approach were used to find special delays for memory polynomial model and found the solution for 20 MHz bandwidth signal. In our research, there were suggested the idea of using the Temporal Pattern Attentio approach with low parameters models Instant Gate Recurrent neural network(IGRNN) and Instant Gate Implict Recurrent neural network (IGIRNN), which was based on the GRU cells [14].

2   Idea Description

2.1   Attention Mechanism

Originally, attention mechanism, for recurrent neural networks, is the correlation process of RNN output vector and hidden vectors, which produced during the sequence processing. Because of each hidden element, contains special information of corresponding element of sequence in process of correlation we can understand how all of hidden elements influence to the output vector. The process of attention mechanism can be divided into three main parts:

1. Process of dot product between hidden vectors of elements and output vector of model, to check the cosine similarity of two vectors.
2. A numerical conversion (i.e., the Softmax [15] 1) is used to numerically convert the correlation produced in the preceding step, in order to normalize and organize the previously computed results into a probability distribution with the total of all element weights equal to 1. Furthermore, by utilizing the Softmax's intrinsic mechanism, the weight of the vital parts may be made more obvious. The Softmax and the coefficients have a connection.

$$\alpha_t = Softmax(h_y, h_t) = \frac{\exp(h_y \cdot h_t)}{\sum_{k=1}^{M} \exp(h_y \cdot h_k)} \tag{1}$$

3. Summarize the hidden vectors multiplied by the significantly coefficients from Softmax function processing, and concatinate it to the output of RNN function (Fig. 3).

$$v = \sum_{k=1}^{M} \alpha_k * h_k \tag{2}$$

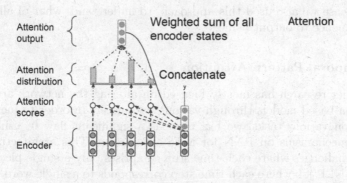

Fig. 3. Description of attention mechanism

and Each if them contain, mostly which contains information of all elements of sequences.

2.2   Memory Term Reduction Approach for DPD

In previous research [21] authors realized attention mechanism for DPD case based on LSTM [10] functions. They tried to found special time delays for GMP [17] Model. In their case there were realized "sequence-to-one" regression task with RNN approach. Attention-LSTM architecture was optimized with SGD-like methods over 1000 epochs. After the process of optimization, the results of $Softmax$ function had some regions of matrix, which were higher, than another parts. After the ensemble average along the diagonals of matrix α, this result present the delays of the memory terms, which is really important for output signal (Fig. 4).

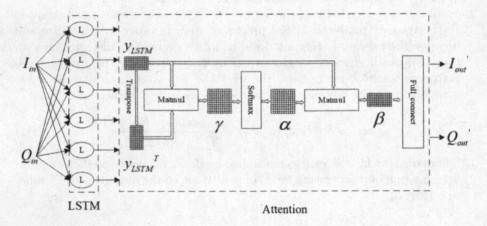

Fig. 4. Description of implementation attention mechanism for finding memory terms delays

In our case we also used this approach, to understand, what of all elements mostly influence to output value.

2.3   Temporal Pattern Attention

While earlier research has mostly focused on altering the network architecture of attention-based models through various parameters in order to increase performance on various tasks, we feel there is a significant flaw in using normal attention mechanisms on RNN for MTS forecasting. This architecture is well suited to challenges where each time step comprises only a single piece of data, such as an NLP job where each time step corresponds to a single word. It fails to disregard variables that are noisy in terms of predicting usefulness when there are several variables in each time step. Furthermore, because the conventional attention mechanism averages information across numerous time steps, it fails to recognize temporal patterns that might be used to forecast.

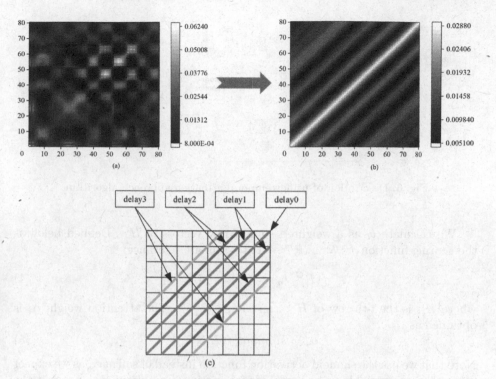

Fig. 5. (a) α matrix of attention block, which were achieved in 1000 epochs of optimization (b) result of ensemble average process of αmatrix (c) delays diagram, based to (b) results

Figure 6 depicts a high-level overview of the suggested paradigm. Given prior RNN hidden states $H \in \mathbb{R}^{m \times (t-1)}$ (where m- count of element per every hidden vector, t- count of element in sequence), the suggested attention mechanism simply attends to its row vectors in the proposed method. Rows with attention weights choose factors that are useful for predicting. The context vector v_t now contains temporal information since it is the weighted sum of the row vectors holding information from various time steps (Fig. 5).

CNN's success lies in no small part to its ability to capture various important signal patterns; as such we use a CNN to enhance the learning ability of the model by applying CNN filters on the row vectors of H. Specifically, we have k filters $C_i \in \mathbb{R}^{1 \times T}$, where T is the maximum length we are paying attention to. If unspecified, we assume $T = w$. Convolutional operations yield $H^C \in \mathbb{R}^{n \times k}$ where $H^C_{i,j}$ represents the convolutional value of the i-th row vector and the j-th filter. Formally, this operation is given by

$$H^C_{i,j} = \sum_{l=1}^{w} H_{i,(t-w-1+l)} x C_{j,T-w+l} \qquad (3)$$

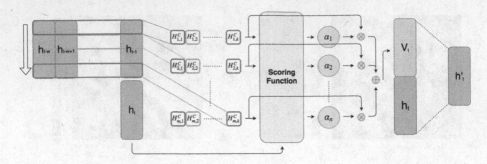

Fig. 6. Description of using temporal attention approach algorithm

We calculate v_t as a weighted sum of row vectors of H^C. Defined below is the scoring function $f : \mathcal{R}^k \times \mathbb{R}^m \to \mathbb{R}$ to evaluate relevance:

$$f(H_i^C, h_t) = (H_i^C)^T \cdot W_\alpha \cdot h_t \tag{4}$$

where H_i^C is the i-th row of H^C, and $W_\alpha \in \mathbb{R}^{k \times m}$. The attention weight α_i is obtained as

$$\alpha_i = sigmoid(f(H_i^C, h_t)) \tag{5}$$

Note that we use the sigmoid activation function instead of softmax, as we expect more than one variable to be useful for forecasting. Completing the process, the row vectors of H^C are weighted by α_i to obtain the context vector $v_t \in \mathbb{R}^k$ 2 Than we concatenate vectors v and h to get the h' to compare them with y signal [19].

2.4   Recurrent Neural Networks for DPD

According to the review of previous research [14], there were designed special recurrent neural networks, which were mainly based on the nonlinear physical characteristics of RF-PAs.

Instant Gated Recurrent Neural Network. There were desined special neural network structure which is more in line with PA characteristic. This structure had two state control units based on current input information. The i-th hidden neuron forward propagation equation presented below (Fig. 7):

$$z_t^i = \sigma(W_z^i x_t + b_z) \tag{6}$$

$$r_t^i = \sigma(W_r^i x_t + b_r) \tag{7}$$

$$\bar{h}_t^i = tanh(W_h^i x_t + U_h^i(r_t^i \otimes h_{t-1}^i) + b_h^i \tag{8}$$

$$h_t^i = (1 - z_t^i) \otimes h_{t-1}^i + z_t^i \otimes \bar{h}_t^i \tag{9}$$

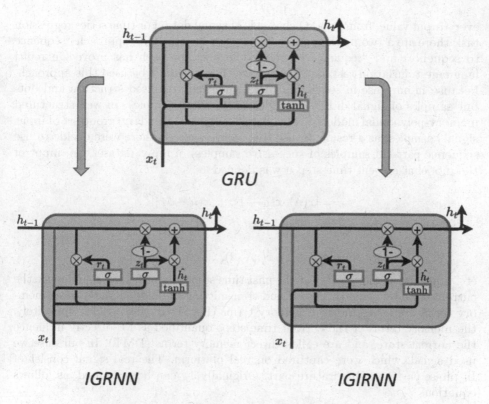

Fig. 7. Schemes of recurrent neural networks

Instant Gated Implict Recurrent Neural Network. To decrease power requirements of digital predistortion, there were designed another neural network, based on IGRNN mode. In this function also presented two gate control units, but information, which is controlled by this gates, gets only from previous stage

$$\bar{h}_t^i = tanh(U_h^i(r_t^i \otimes h_{t-1}^i) + b_h^i \tag{10}$$

As a result, this form of network topology is more in accordance with the properties of a power amplifier, allowing it to have outstanding expression ability while being computationally simple. The two state control units established here only take the current input information x_t in the new structure.

2.5 Behavioral Modeling of TPA Approach Based on IGRNN or IGIRNN

To solve the regression case for time series task with recurrent approach, there is not necessary to use special embedding, for every input token. In our case we also didn't use it, because of we couldn't generate informative embedding for every input value, from I and Q channels of signal data. For time series regression task there are a two main approaches to solve it with RNN approach "sequence to sequence" and "sequence to one". The first approach has proved in multi language translation task but in our case there couldn't be used this approach, because in our case in sequences there isn't well organised structure and output samples of signal didn't depended on previous samples. In well structured memory-polynomial models, every output samples, depends on group set of input signals samples, as a result, to use this memory affect, there were decide to use sequence part (M samples of successive samples) of input dataset. So, input of the model at current time step n was defined as.

$$x_n = [x(n), x(n-1), ..., x(n-M)]^T,$$

for the time step n-1 we should take

$$x_{n-1} = [x(n-1), x(n-2), ..., x(n-M-1)]^T$$

etc. Since the memory terms at the past time steps, quantified as T, influence the current state by being processed and transmitting the hidden state, these memory terms are called indirect memory terms (IDMT) in this article. Oppositely, the memory terms at the current time steps quantified as M, directly influence the current state, and are called direct memory terms (DMT). In our case we used signal, which were captured on real platform. The real signal consist of In phase part and Quadrature part. Originally x_n can be presented, as follows equation

$$x_n = i_n + j \cdot q_n \tag{11}$$

so, the batch variant of input sequence will be presented as:

$$I_n = \begin{bmatrix} i_n & i_{n-1} & \cdots & i_{n-M} \\ \vdots & \ddots & & \vdots \\ i_{n-T} & i_{n-1-T} & \cdots & i_{n-M-T} \end{bmatrix} \tag{12}$$

and

$$Q_n = \begin{bmatrix} q_n & q_{n-1} & \cdots & q_{n-M} \\ \vdots & \ddots & & \vdots \\ q_{n-T} & q_{n-1-T} & \cdots & q_{n-M-T} \end{bmatrix} \tag{13}$$

According the ides, to use all information from the previous samples of sequences more effectively(IDMT should work better) there were decided to use attention approach. The idea of using some vectors from processing of previous samples of model allowed to find some hidden patterns and hidden dependence's for output samples. Attention idea allows to accumulate some hidden data information of whole sequence samples and realize the idea of delay term memory. The conventional attention mechanism averages information across numerous, it fails to recognize temporal patterns that might be used to linearization of RF PAs output. At the Fig. 9 you can see the idea of using TPA Approach for DPD task. More detail explanation off this idea you can see upper Sect. 2.3 (Fig. 8).

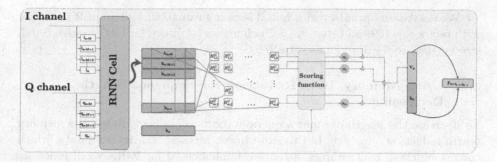

Fig. 8. Scheme of TPA approach for DPD model

In our case we realize temporal pattern Attention approach, which is based on IGIRNN 2.4 and IGRNN 2.4 recurrent cell functions.

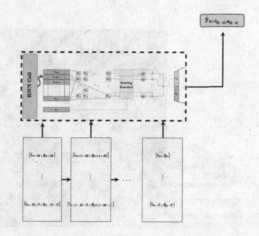

Fig. 9. Scheme of sequance to one regression task

2.6   Validation Metrics

To assess the quality of the solution obtained as a result of the optimization of this loss functional, we will further use the normalized mean square error quality metric, measured in decibels:

$$\text{NMSE}(y,\overline{y}) := 10\log_{10}\left\{\frac{\sum_{k=1}^{m}(y_k - \overline{y}_k)^2}{\sum_{k=1}^{m}x_k^2}\right\} \quad \text{dB}.$$

3   Experiments

In the following experiments we use 2 different training signals: 80 MHz and 200 MHz. We split them into 2 parts. 80% of signal is used for training, and other 20% is used for testing. In Table 1 we list experiments on TPA approach.

We use Adam optimizer with initial learning rate 0.001 and StepLR scheduler with factor $\gamma = 0.99$ and step size 5. Each model is trained for 1500 epochs. Batch size is equal to 64. Random seed is fixed.

3.1   Using Memory Term Reduction for Sequence Length Decreasing

To decrease the length of input sequences there were decide to use the memory tearm reduction approach. In this experiment we used 80 Mhz band data, which consist of 99840 I, Q samples, attention implemented for IGRNN cell function. After 1000 epochs of training and calculating of ensemble average for each diagonals of α matrix, there were got the next matrix, described delays of sequence:

Results in Table 1 show that TPA improves NMSE for about 1 dB for LSTM and GRU, while increase in number of parameters is negligible. IGRNN performs better that GRU with equal amount of parameters (Fig. 10).

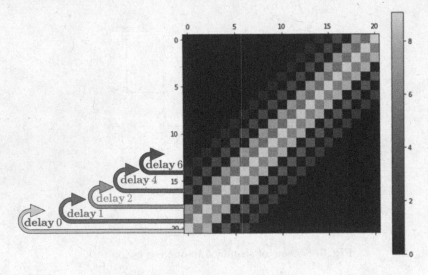

Fig. 10. Matrix with memory depth of 20; The right-side vertical scale shows the α values

This approach allowed us to decrease architecture since $16000 \pm 500\,to$ 15000 ± 500 and performance decrease since -42.4 dB to -41.4 dB. So, this approach allowed us to decrease number of parameters and achive not really bad performance also, this approach allows us to keep less count of element every iteration of optimization and as a result it allows us to optimize model faster.

3.2   Using TPA Approach

To increase lingering ability of DPD model architecture, there were realized TPA approach, which were described at the previous Subsect. 2.3, 2.5. In this experiments we used two types of signals with 80 MHz and 200 MHz bandwidth. In this type of experiments we also used Bidirectional technique, which often used for time series models. According the results, which were presented in Table 1, the best performance were achieved by LSTM architecture on all signals, but twice less count of model parameters (IGIRNN) achieved performance near the sames, as performance of LSTM. For example the best result on 80 MHZ bandwidth signal were achieved with LSTM and it's score is −42.58 dB with 24579 parameters, but −42.03 dB were achieved by IGRNN model with 13619 parameters (Figs. 11, 12, 13 and 14).

Table 1. Experiments on Temporal Pattern Attention. $w = 11$, $k = 9$ and $m = 20$ for all experiments below

rnn_name	bidirectional	num_layers	type_(Mhz)	attention	nmse	num_params
gru	-	2	80	-	-38.64	4000
gru	-	2	80	+	-37.39	4859
gru	+	3	80	-	-40.85	17800
gru	+	3	80	+	-42.4	18659
igirnn	-	2	80	-	-35.02	1840
igirnn	-	2	80	+	-34.86	2699
igirnn	+	3	80	-	-37.19	9360
igirnn	+	3	80	+	-37.25	10219
igrnn	-	2	80	-	-37.65	2320
igrnn	-	2	80	+	-37.69	3179
igrnn	+	3	80	-	-40.5	12760
igrnn	+	3	80	+	-42.03	13619
lstm	-	2	80	-	-39.68	5320
lstm	-	2	80	+	-38.15	6179
lstm	+	3	80	-	-41.65	23720
lstm	+	3	80	+	-42.58	24579
gru	-	2	200	+	-34.71	4000
gru	-	2	200	-	-35.35	4859
gru	+	3	200	+	-38.04	17800
gru	+	3	200	-	-39.32	18659
igirnn	-	2	200	+	-31.31	1840
igirnn	-	2	200	-	-32.93	2699
igirnn	+	3	200	+	-35.15	9360
igirnn	+	3	200	-	-36.38	10219
igrnn	-	2	200	+	-33.28	2320
igrnn	-	2	200	-	-34.50	3179
igrnn	+	3	200	+	-38.03	12760
igrnn	+	3	200	-	-38.46	13619
lstm	-	2	200	+	-35.59	5320
lstm	-	2	200	-	-36.04	6179
lstm	+	3	200	+	-38.75	23720
lstm	+	3	200	-	-39.51	24579

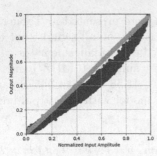

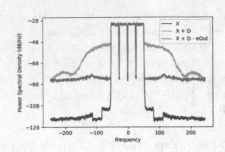

Fig. 11. Amplitude input - amplitude output plot with and without DPD for 80 Mhz OFDM signal

Fig. 12. Power spectrum density of 80 MHz spectrum bandwidth signal dataset

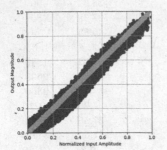

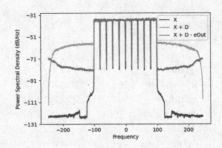

Fig. 13. "Amplitude input - amplitude output" plot with and without DPD for 200 MHz OFDM signal

Fig. 14. Power spectrum density of 200 MHz spectrum bandwidth signal dataset

4   Conclusion

The use of low-complexity neural networks, based on the RNN model with an attention mechanism, is offered as a unique technique. The two models offered to overcome the problem of vanishing gradients in RNN model training differ from the classic RNN model due to their personalized attention strategy. As a result, they are better able to characterize the short-term and long-term memory effects of RF-PA. We compared the custom attention mechanism (temporal memory approach) to other RNN-based variant models, such as GRU and LSTM models, during the course of our research, and as a result, the custom attention mechanism (temporal memory approach) allows us to show simpler models high performance and can effectively reduce model parameters. Theoretical research and practical data indicate that the Temporal pattern method with IGRNN has superior performance and much reduced computing complexity when compared to RNN variant models such as the GRU and LSTM models, which have comparable performance.

References

1. Bengio, Y., Simard, P., Frasconi, P.: Learning long-term dependencies with gradient descent is difficult. IEEE Trans. Neural Netw. **5**(2), 157–166 (1994)
2. Briffa, M.A.: Linearization of RF power amplifiers. Ph.D. thesis, Victoria University (1996)
3. Chen, P., Alsahali, S., Alt, A., Lees, J., Tasker, P.J.: Behavioral modeling of GAN power amplifiers using long short-term memory networks. In: 2018 International Workshop on Integrated Nonlinear Microwave and Millimetre-Wave Circuits (INM-MIC), pp. 1–3. IEEE (2018)
4. Chung, J., Gulcehre, C., Cho, K., Bengio, Y.: Empirical evaluation of gated recurrent neural networks on sequence modeling. arXiv preprint arXiv:1412.3555 (2014)
5. Ding, L., et al.: A robust digital baseband predistorter constructed using memory polynomials. IEEE Trans. Commun. **52**(1), 159–165 (2004)
6. Ghannouchi, F.M., Hammi, O., Helaoui, M.: Behavioral Modeling and Predistortion of Wideband Wireless Transmitters. Wiley, Hoboken (2015)
7. Guan, L., Zhu, A.: Green communications: digital predistortion for wideband RF power amplifiers. IEEE Microwave Mag. **15**(7), 84–99 (2014)
8. Hadi, M.U., Awais, M., Raza, M.: Multiband 5G NR-over-fiber system using analog front haul. In: 2020 International Topical Meeting on Microwave Photonics (MWP), pp. 136–139. IEEE (2020)
9. Hochreiter, S.: The vanishing gradient problem during learning recurrent neural nets and problem solutions. Internat. J. Uncertain. Fuzziness Know.-Based Syst. **6**(02), 107–116 (1998)
10. Hochreiter, S., Schmidhuber, J.: Long short-term memory. Neural Comput. **9**(8), 1735–1780 (1997)
11. Hornik, K.: Approximation capabilities of multilayer feedforward networks. Neural Netw. **4**(2), 251–257 (1991)
12. Kang, S., Sung, E.T., Hong, S.: Dynamic feedback linearizer of RF CMOS power amplifier. IEEE Microwave Wirel. Compon. Lett. **28**(10), 915–917 (2018)
13. Katz, A., Wood, J., Chokola, D.: The evolution of PA linearization: from classic feedforward and feedback through analog and digital predistortion. IEEE Microw. Mag. **17**(2), 32–40 (2016)
14. Li, G., Zhang, Y., Li, H., Qiao, W., Liu, F.: Instant gated recurrent neural network behavioral model for digital predistortion of RF power amplifiers. IEEE Access **8**, 67474–67483 (2020)
15. Luong, M.T., Pham, H., Manning, C.D.: Effective approaches to attention-based neural machine translation. arXiv preprint arXiv:1508.04025 (2015)
16. Luongvinh, D., Kwon, Y.: Behavioral modeling of power amplifiers using fully recurrent neural networks. In: 2005 IEEE MTT-S International Microwave Symposium Digest, pp. 1979–1982. IEEE (2005)
17. Morgan, D.R., Ma, Z., Kim, J., Zierdt, M.G., Pastalan, J.: A generalized memory polynomial model for digital predistortion of RF power amplifiers. IEEE Trans. Signal Process. **54**(10), 3852–3860 (2006)
18. Qin, Y., Song, D., Chen, H., Cheng, W., Jiang, G., Cottrell, G.: A dual-stage attention-based recurrent neural network for time series prediction. arXiv preprint arXiv:1704.02971 (2017)
19. Shih, S.-Y., Sun, F.-K., Lee, H.: Temporal pattern attention for multivariate time series forecasting. https://doi.org/10.48550/arxiv.1809.04206. https://arxiv.org/abs/1809.04206

20. Wang, D., Aziz, M., Helaoui, M., Ghannouchi, F.M.: Augmented real-valued time-delay neural network for compensation of distortions and impairments in wireless transmitters. IEEE Trans. Neural Netw. Learn. Syst. **30**(1), 242–254 (2018)
21. Yu, H., Xu, G., Liu, T., Huang, J., Zhang, X.: A memory term reduction approach for digital pre-distortion using the attention mechanism. IEEE Access **7**, 38185–38194 (2019)
22. Zhang, J.S., Xiao, X.C.: Predicting chaotic time series using recurrent neural network. Chin. Phys. Lett. **17**(2), 88–90 (2000). https://doi.org/10.1088/0256-307x/17/2/004
23. Zhu, A.: Decomposed vector rotation-based behavioral modeling for digital predistortion of RF power amplifiers. IEEE Trans. Microw. Theory Tech. **63**(2), 737–744 (2015)
24. Zhu, A., Pedro, J.C., Brazil, T.J.: Dynamic deviation reduction-based Volterra behavioral modeling of RF power amplifiers. IEEE Trans. Microw. Theory Tech. **54**(12), 4323–4332 (2006)

Forecasting with Using Quasilinear Recurrence Equation

Anatoly Panyukov[ID], Tatiana Makarovskikh[✉][ID], and Mostafa Abotaleb[ID]

South Ural State University, Chelyabinsk, Russia
{paniukovav,Makarovskikh.T.A}@susu.ru
http://www.susu.ru

Abstract. We developed a new approach to the analysis of time series based on the use of quasi-linear recurrence relations. Unlike neural networks, this approach makes it possible to explicitly obtain high-quality quasi-linear difference equations (adequately describing the considered process). Currently, we developed and tested methods for identifying the parameters of a single equation. The research considers the identification algorithm for parameters of quasilinear recurrence equation. We use it to solve the problem of regression analysis with mutually dependent observable variables, which allows to implement the generalized last deviations method (GLDM). Using this model we held the computational experiment. The model using the identified parameters allows to obtain the long-time forecast.

Keywords: Forecasting · Time series · Quasilinear model · Generalized least deviations method

1 Introduction

Currently, a lot of experience has been accumulated in measuring vibration signals, developing methods of vibration diagnostics and forecasting the condition and resource of mechanical systems. In this regard, one of the most urgent directions is to improve the accuracy and speed of determining diagnostic signs. The above applies, first of all, to unique, highly loaded mechanical systems considered, for example, by [10,12]. The solution to this problem in many cases can be obtained through the dynamic characteristics of mechanical systems. The determination of these characteristics is greatly facilitated by the correct choice of a diagnostic mathematical model that establishes a connection between the space of object states and the space of diagnostic features. Dynamic models presented in the form of difference equations, phenomenological, structural, regression models, etc. are considered as them. The choice of a particular model depends on the defined characteristics and the nature of the analyzed process.

The work was supported by Act 211 Government of the Russian Federation, contract No. 02.A03.21.0011. The work was supported by the Ministry of Science and Higher Education of the Russian Federation (government order FENU-2020-0022).

N. Olenev et al. (Eds.): OPTIMA 2022, CCIS 1739, pp. 183–195, 2022.
https://doi.org/10.1007/978-3-031-22990-9_13

Identification using different statistical methods, neural networks, or mathematical models has been urgent for a long time in different spheres of life. Nowadays these methods are used not only in industry but also for attempts to forecast the development of Covid-19 pandemics. For example, [2] compares the quality of forecasting the pandemics process by different well known models, develops the software running all of these methods and holds the computational experiments using Covid-19 time series. The authors conclude that their forecasting system can be implemented to any kind of time series. The most of forecasts, especially for big data are made using different neural network models. For example, [1,11] considers the neural network model that can be used to forecast changes in the price of ferrosilicon in the domestic market of the Russian Federation in the short term. His model is distinguished by high forecasting accuracy and can be useful in substantiating strategic decisions in the activities of branch research institutes and metallurgical enterprises. [5,13] describes the econometric models of the qualitative economic indicator of metallurgical branch, production applicable for estimation of the statistical features of production ferrous metals and perspective development ferrous metallurgy. Nevertheless, all such models look like a black magic box, allowing to obtain some appropriate answer for some input data. Some researchers are using so-called cognitive modelling to increase the quality of forecasting by neural networks. The paper be [3] aims to compare the performance of cognitive and mathematical time series predictors, regarding accuracy. The authors discover that the results of their experiment showed that the cognitive models have at least equivalent accuracy in comparison to the ARIMA models. Most of these approaches are used for forecasting some economical units like production volume, some logistics parameters etc.

Since all the listed above models work for short time forecasting the task of developing mathematical approach that makes it possible to explicitly obtain high-quality quasi-linear difference equations (adequately describing the considered process) is urgent. There are known some researches in this field like [5], where the proposed model includes data cleaning, data smoothing and final data after preprocessing fed into regression-based model to predict industrial electric power consumption. But this paper as lots of others again considers statistical methods.

In our paper we discuss our methods for identifying the parameters of a single equation. The research considers the identification algorithm for parameters of quasilinear recurrence equation. We use it to solve the problem of regression analysis with mutually dependent observable variables, which allows to implement the generalized last deviations method (GLDM). Using this model we held the computational experiment. The model using the identified parameters allows to obtain the long-time forecast. Unlike neural networks, this approach makes it possible to explicitly obtain high-quality quasi-linear difference equations (adequately describing the considered process).

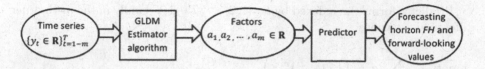

Fig. 1. The scheme of the model implementation

In this paper we try to implement the considered by [7] algorithms to determine the coefficients $a_1, a_2, a_3 \ldots, a_m \in \mathbb{R}$ of a m-th order quasilinear autoregressive model

$$y_t = \sum_{j=1}^{n(m)} a_j g_j(\{y_{t-k}\}_{k=1}^m) + \varepsilon_t, \quad t = 1, 2, \ldots, T \tag{1}$$

by up-to-date information about of values of state variables $\{y_t \in \mathbb{R}\}_{t=1-m}^T$ at time instants t; here $g_j : (\{y_{t-k}\}_{k=1}^m) \to \mathbb{R}$, j=1,2,...n(m) are given $n(m)$ functions, and $\{\varepsilon_t \in \mathbb{R}\}_{t=1}^T$ are unknown errors.

2 Notation and Statement of the Problem

The considered algorithm works like the following (Fig. 1). GLDM-algorithm [7] gets a time series $\{y_t \in \mathbb{R}\}_{t=-1-m}^T$ of length $T + m \geq (1 + 3m + m^2)$ as an input data and determines the factors $a_1, a_2, a_3 \ldots, a_m \in \mathbb{R}$ by solving the optimization problem

$$\sum_{t=1}^T \arctan \left| \sum_{j=1}^{n(m)} a_j g_j(\{y_{t-k}\}_{k=1}^m) - y_t \right| \to \min_{\{a_j\}_{j-1}^{n(m)} \subset \mathbb{R}} \tag{2}$$

The Cauchy distribution

$$F(\xi) = \frac{1}{\pi} \arctan(\xi) + \frac{1}{2}$$

has the maximum entropy among distributions of random variables that have no mathematical expectation and variance. That's why function $\arctan(*)$ is applied for this research.

Further we consider an m-th order model with quadratic nonlinearity, therefore the basic set of $g_{(*)}$ functions contains

$$g_{(k)}(\{y_{t-k}\}_{k=1}^m) = y_{t-k},$$
$$g_{(kl)}(\{y_{t-k}\}_{k=1}^m) = y_{t-k} \cdot y_{t-l},$$
$$k = 1, 2, \ldots, m; \ l = k, k+1, \ldots, m.$$

Obviously, in this case $n(m) = 2m + C_m^2 = m(m+3)/2$, and the numbering of $g_{(*)}$ functions can be arbitrary.

Predictor forms the indexed by $t = 1, 2, \ldots, T-1, T$ family of the m-th order difference equations

$$\overline{y[t]}_\tau = \sum_{j=1}^{n(m)} a_j^* g_j \left(\{\overline{y[t]}_{\tau-k}\}_{k=1}^m \right),$$

$$\tau = t, t+1, t+2, t+3, \ldots, T-1, T, T+1, \ldots \quad (3)$$

for lattice functions $\overline{y[t]}$ with values $\overline{y[t]}_\tau$ which interpreted as constructed at time moment t the forecasts for y_τ. Let us use the solution of the Cauchy problem for its difference equation (3) under the initial conditions

$$\overline{y[t]}_{t-1} = y_{t-1}, \; \overline{y[t]}_{t-2} = y_{t-2}, \; \ldots, \; \overline{y[t]}_{t-m} = y_{t-m},$$

$$t = 1, 2, \ldots, T-1, T \quad (4)$$

to find the values of the function $\overline{y[t]}$.

So we have the set $\overline{Y}_\tau = \left\{ \overline{y[t]}_\tau \right\}_{t=1}^T$ of possible prediction values of y_τ. Further we use this set to estimate the probabilistic characteristics of the y_τ value.

3   Evaluating by GLDM

Problem (2), i.e. problem of GLDM-estimation, is a multi-extremal optimization problem. GLDM-estimates are robust to the presence of a correlation of values in $\{y_t \in \mathbb{R}\}_{t=-1-m}^T$, and (with appropriate settings) are the best for probability distributions of errors with heavier (than normal distribution) tails (see [10]). All the above shows the feasibility of solving the identification problem (1) with usage solution (2).

Let us use the interrelation between GLDM-estimates and estimates by the weighted least deviation method considered by [4] (WLDM-estimates) to solve problems (2) of higher dimension.

Let us consider the algorithm of GLDM estimation (see [6]) in terms of this paper. First of all let us consider WLDM estimation algorithm used in GLDM algorithm.

3.1   Evaluating by WLDM

Algorithm WLDM-estimator [7] gets a time series $\{y_t \in \mathbb{R}\}_{t=1-m}^T$ and weight factors $\{p_t \in \mathbb{R}^+\}_{t=1}^T$ as an input data and calculates the factors

$$a_1, a_2, a_3 \ldots, a_{n(m)} \in \mathbb{R}$$

by solving the optimization problem

$$\sum_{t=1}^T p_t \cdot \left| \sum_{j=1}^{n(m)} a_j g_j(\{y_{t-k}\}_{k=1}^m) - y_t \right| \longrightarrow \min_{\{a_j\}_{j=1}^{n(m)} \in \mathbb{R}^{n(m)}} \quad (5)$$

This problem represents the problem of convex piecewise linear optimization, and the introduction of additional variables reduces it to the problem of linear programming

$$\sum_{t=1}^{T} p_t z_t \to \min_{\substack{(a_1,a_2,\ldots,a_{n(m)})\in\mathbb{R}^m, \\ (z_1,z_2,\ldots,z_T)\in\mathbb{R}^T}}, \tag{6}$$

$$-z_t \leq \sum_{j=1}^{n(m)} [a_j g_j(\{y_{t-k}\}_{k=1}^m)] - y_t \leq z_t, \tag{7}$$

$$z_t \geq 0, \qquad t = 1, 2, \ldots, T \tag{8}$$

Problem 6–8 has a canonical form with $n(m) + T$ variables and $3n$ inequality constraints including the conditions for the non-negativity of the variables z_j, $j = 1, 2, \ldots, T$. The dual task to task (6) is

$$\sum_{t=1}^{T} (u_t - v_t) y_t \to \max_{u,v\in\mathbb{R}^T}, \tag{9}$$

$$\sum_{t=1}^{T} a_j g_j(\{y_{t-k}\}_{k=1}^m)(u_t - v_t) = 0, \ j = 1, 2, \ldots, n(m), \tag{10}$$

$$u_t + v_t = p_t, \quad u_t, v_t \geq 0, \quad t = 1, 2, \ldots, T. \tag{11}$$

Let us introduce variables $w_t = u_t - v_t$, $t = 1, 2, \ldots, T$. Conditions (11) imply

$$u_t = \frac{p_t + w_t}{2}, \quad v_t = \frac{p_t - w_t}{2}, \quad -p_t \leq w_t \leq p_t, \quad t = 1, 2, \ldots, T. \tag{12}$$

Therefore the optimal value of the problem (9)–(11) is equal to the optimal value of problem

$$\sum_{t=1}^{T} w_t \cdot y_t \to \max_{w\in\mathbb{R}^T}, \tag{13}$$

$$\sum_{t=1}^{T} g_j(\{y_{t-k}\}_{k=1}^m) \cdot w_t = 0, \ j = 1, 2, \ldots, n(m), \tag{14}$$

$$-p_t \leq w_t \leq p_t, \ t = 1, 2, \ldots, T. \tag{15}$$

Constraints (14) define a $(T - n(m))$-dimensional linear subspace $\mathcal{L}$ with $(n(m) \times T)$-matrix

$$S = \begin{bmatrix} g_1(\{y_{1-k}\}_{k=1}^m) & g_1(\{y_{2-k}\}_{k=1}^m) & \cdots & g_1(\{y_{T+1-k}\}_{k=1}^m) \\ g_2(\{y_{1-k}\}_{k=1}^m) & g_2(\{y_{2-k}\}_{k=1}^m) & \cdots & g_2(\{y_{T+1-k}\}_{k=1}^m) \\ \vdots & \vdots & \ddots & \vdots \\ g_{n(m)}(\{y_{1-k}\}_{k=1}^m) & g_{n(m)}(\{y_{1-k}\}_{k=1}^m) & \cdots & g_{n(m)}(\{y_{1-k}\}_{k=1}^m) \end{bmatrix},$$

constraints (15) define T-dimensional brick $\mathcal{T}$.

A simple structure of the allowable set of problem (13)–(15): the intersection of $(T - n(m))$- dimensional linear subspace $\mathcal{L}$ (14) and T - dimensional brick $\mathcal{T}$ (15), – allow us to find its solution by an algorithm uses the gradient projection of the goal function (13) (i.e. vector $\nabla = \{y_t\}_{t=1}^T$) on an acceptable area $\mathcal{L} \cap \mathcal{T}$ that is defined by restrictions (14)–(15). The matrix of the projection operator on $\mathcal{L}$ is

$$S_{\mathcal{L}} = E - S^T \cdot \left(S \cdot S^T\right)^{-1} \cdot S,$$

and gradient projection on $\mathcal{L}$ is $\nabla_{\mathcal{L}} = S_{\mathcal{L}} \cdot \nabla$. Besides, if the external normal on some face of the brick forms an acute angle with a gradient projection $\nabla_{\mathcal{L}}$ the movement along this face is zero.

DualWLDMSoluter. Algorithm 1 to solve problem (13)-(15) begins the search of the optimal solution at 0, moving along direction $\nabla_{\mathcal{L}}$. If the current point falls on the face of brick $\mathcal{T}$, then the corresponding coordinate in the direction of the moving is assumed to be 0.

Algorithm 1. DualWLDMSoluter

Require: :

∇_L ▷ Gradient projection

$\{p_t \in \mathbb{R}^+\}_{t=1}^T$ ▷ Weight factors

Ensure: :

$$w^* = \arg \max_{w \in R^T} \sum_{i=1}^T w_i \cdot y_i \qquad \qquad \text{▷ Optimal dual solution}$$

$$R^* = \{t \in T : \; |w_7 4 t^*| = p_t\} \qquad \qquad \text{▷ Active restrictions}$$

1: $w \leftarrow \{w_i = 0: \; i = 1, 2, \ldots, T\}; \; R \leftarrow \emptyset; \; g = \nabla_L$
2: **while** $(\alpha_* \neq 0)$ **do**
3:      $\{(\alpha_*, t_*) \leftarrow \arg\max \{\alpha \geq 0: \; -p_t \leq w_t + \alpha g_t \leq p_t\}\}$
4:      $w \leftarrow w + \alpha_* g; \; g_{t_*} \leftarrow 0; \; R := R \cup \{t_*\};$
5: **end while**
6:   $w^* = w, \; R^* = R$
    return (w^*, R^*)

Computational complexity of such algorithm does not exceed $O(T^2)$ due to the simple structure of the admissible set: intersection of T-dimensional cuboid (15) and $(T - n(m))$-dimensional linear subspace (14).

If (w^*, R^*) is the result of executing the Algorithm 1, then w^* is the optimal solution to the problem (13)–(15), and the optimal solution of the problem (9)–(11) is equal to

$$u_t^* = \frac{p_t + w_t^*}{2}, \quad v_t^* = \frac{p_t - w_t^*}{2}, \quad t = 1, 2, \ldots, T.$$

It is following from the complementarity condition for a pair of mutually dual problems (6)–(8) and (9)–(11) that

$$y_t = \sum_{j=1}^{n(m)} [a_j g_j(\{y_{t-k}\}_{k=1}^m)] \qquad \forall t \notin R^*, \tag{16}$$

$$y_t = \sum_{j=1}^{n(m)} [a_j g_j(\{y_{t-k}\}_{k=1}^m)] + z_t^*, \quad \forall t \in R^* : w_t^* = p_t, \tag{17}$$

$$y_t = \sum_{j=1}^{n(m)} [a_j g_j(\{y_{t-k}\}_{k=1}^m)] - z_t^*, \quad \forall t \in R^* : w_t^* = -p_t. \tag{18}$$

In fact, the solution $(\{a_j^*\}_{j-1}^{n(m)}, z^*)$ of linear algebraic equations system (16)–(18) represents the dual optimal solution of problem (13)–(15) and the optimal solution of the problem (5), that proves the validity of the following theorem.

Theorem 1. *Let w^* be the optimal solution of the problem, (13)–(15), Let $(\{a_j^*\}_{j-1}^{n(m)}, z^*)$ be solution of a system of linear algebraic equations (16)–(18). Then $(\{a_j^*\}_{j-1}^{n(m)}$ is the optimal solution to the problem (5).*

The above allows us to propose WLDM-estimator Algorithm 2. The main problem with the use of WLDM-estimator is the absence of general formal rules for choosing weight coefficients. Consequently, this approach requires additional research.

Algorithm 2. WLDM-estimator

Require: :
  $S = \{S_t \in \mathbb{R}^N\}_{t \in T}$ ▷ The matrix of a linear subspace $\mathcal{L}$
  $\nabla_{\mathcal{L}}$ ▷ Gradient projection on $\mathcal{L}$
  $\{p_t \in \mathbb{R}^+\}_{t=1}^T$ ▷ Weight factors
  $\{y_t \in \mathbb{R}^+\}_{t=1-m}^T$ ▷ Values of the given state variables
Ensure: :
  $A^* \in \mathbb{R}^{n(m)}$ ▷ Optimal primal solution
  $z^* \in \mathbb{R}^T$ ▷ Restrictions

1: $(w^*, R^*) \leftarrow$ **DualWLDMSoluter** $(\nabla_{\mathcal{L}}, \quad \{p_t \in \mathbb{R}^+\}_{t=1}^T)$
2: $S^* \leftarrow \{S_t : t \notin R^*\}; \ y^* \leftarrow \{y_t : t \notin R^*\}$ ▷ System (16) matrix
3: $(A^*)^\mathsf{T} \leftarrow y^\mathsf{T} \cdot (S^*)^{-1}$ ▷ System (16) solution
4: $z^* \leftarrow (A^*)^\mathsf{T} S - y$ ▷ Find restrictions
  return (A^*, z^*)

The established in [6,9] results allow us to reduce the problem of determining GLDM estimation to an iterative procedure with WLDM estimates.

3.2 GLDM Estimation Algorithm

Problem (2) of GLDM estimation is a concave optimization problem. GLDM-estimates are robust to the presence of a correlation of values in $\{X_{jt} : t = 1, 2, \ldots, T; \; j = 1, 2, \ldots, N\}$, and (with appropriate settings) like the best for probability distributions of errors with heavier (than normal distribution) tails [10]. The above shows the feasibility of solving the identification problem (1) by Algorithm (2). The established in [6] results allow us to reduce the problem of determining GLDM estimation to an iterative procedure with WLDM estimates (see Algorithm 3).

Algorithm 3. GLDM-estimator

Require: :

   $S = \{S_t \in \mathbb{R}^N\}_{t \in T}$ ▷ The matrix of a linear subspace $\mathcal{L}$

   ∇_L ▷ Gradient projection on $\mathcal{L}$

   $\{p_t \in \mathbb{R}^+\}_{t=1}^T$ ▷ Weight factors

   $\{y_t \in \mathbb{R}^+\}_{t=1-m}^T$ ▷ Values of the given state variables

Ensure: :

   $A^* \in \mathbb{R}^{n(m)}$ ▷ Optimal GLDM solution

   $z^* \in \mathbb{R}^T$ ▷ Residuals

1: $p \leftarrow \{p_t = 1 : t = 1, 2, \ldots, T\}$
2: $(A^{(1)}, z^{(1)}) \leftarrow$
3:    $\leftarrow$ **WLDMSoluter** $\left(S, \nabla_{\mathcal{L}}, \{p_t\}_{t=1}^T, \{y_t\}_{t=1-m}^T\right)$
4: **for all** $(t = 1, 2, \ldots T)$ **do**
5:      $p_t \leftarrow \left(1/\left(1 + (z_t^{(1)})^2\right)\right)$
6: **end for**
7: $(A^{(2)}, z^{(2)}) \leftarrow$ **WLDMSoluter** $\left(S, \nabla_{\mathcal{L}}, \{p_t\}_{t=1}^T, \{y_t\}_{t=1-m}^T\right)$
8: $k \leftarrow 2$
9: **while** $\left(A^{(k)} \neq A^{(k-1)}\right)$ **do**
10:     **for all** $(t = 1, 2, \ldots T)$ **do**
11:        $p_t^{(k)} \leftarrow \left(1/\left(1 + (z_t^{(k)})^2\right)\right)$
12:     **end for**
13:     $((A, z)) \leftarrow$ **WLDMSoluter** $\left(S, \nabla_{\mathcal{L}}, \{p_t^{(k)}\}_{t=1}^T, \{y_t\}_{t=1-m}^T\right)$
14:     $(A^{(k+1)}, z^{(k+1)}) \leftarrow (A, z)$
15:     $k \leftarrow (k+1)$
16: **end while**
17: $z^* \leftarrow z^{(k)}, \quad (A^*) \leftarrow A^{(k)}$ ▷ Find restrictions
    return (A^*, z^*)

Theorem 2. *The sequence* $\{(A^{(k)}, z^{(k)})\}_{k=1}^{\infty}$, *constructed by GLDM-estimator Algorithm, converges to the global minimum* (a^*, z^*) *of the problem* (2).

The description of **GLDM-estimator** Algorithm shows that its computational complexity is proportional to the computational complexity of the algorithm for solving of primal and/or dual WLDM problems (5). Multiply computational experiments show that the average number of iterations of **GLDM-estimator** Algorithm is equal to the number of coefficients in the identified equation. If this hypothesis is true then computational complexity in solving practical problems does not exceed

$$O((n(m))^3 T + n(m) \cdot T^2).$$

It is necessary to take into account that the search and finding of the high-order autoregression equation have their own specific conditions. One of these conditions, in particular, is the high sensitivity of the algorithm to rounding errors. To eliminate the possibility of error in the calculations, it is necessary to accurately perform basic arithmetic operations on the field of rational numbers [8] and supplement them with parallelization.

4 Predictor

Predictor forms the indexed by $t = 1, 2, \ldots, T - 1, T$ family of the m-th order difference equations (3) for lattice functions $\overline{y[t]}$ with values $\overline{y[t]}_\tau$ that interpreted as constructed at time moment t the forecast for y_τ. Let us use the solution of the Cauchy problem for its difference equation (3) under the initial conditions (4) to find the values of the function $\overline{y[t]}$. So we have the set $\overline{Y}_\tau = \left\{ \overline{y[t]}_\tau \right\}_{t=1}^{T}$ of possible prediction values of y_τ. Further we use this set to estimate the probabilistic characteristics of the y_τ value. It should be written as Algorithm 4.

5 Experimental Results

Let us consider the computational experiment on constructing the solution of Cauchy problem to one quasi-linear difference equation, the identification of this equation, and let us show that the obtained solution shows the high quality of the considered algorithm. So, we present the results of computational experiment to solve the unknown recurrence equation of the time series.

We consider the process shown in Fig. 2. This process has data for 655 days. Let us for the experiment use consider the time series satisfying this process with length 150, 300, 500, and 655 days for the model of the second order

$$y_t = (a_1 y_{t-1} + a_2 y_{t-2}) + \left(a_3 y_{t-1}^2 + a_4 y_{t-1} y_{t-2} + a_5 y_{t-2}^2 \right).$$

Hence, algorithm is to define five coefficients $a_1, \ldots, a_5$.

Identification results are presented in Table 1. It shows that the experiment for 300 points gives the lowest value of the loss function; coefficients of the model are significant. Most likely, after increasing the length of the observed time series the experiments on the value of the loss function allow receive the dependency of influence the time series length on its value.

After comparing the results of calculation using our model with machine learning models we have the following (see Table 2). The input data is the time

---

Algorithm 4. Predictor

Require: :
  $Y=\{y_t \in \mathbb{R}^+\}_{t=1-m}^T$ ▷ Values of the given state variables
  $A= \{a_i\}_{i=1}^{n(m)}$ ▷ WLDM solution
Ensure: :
  PY[1:T][1:T]: $PY[t][\tau] = \overline{y[t]}_\tau$ ▷ forecast for y_τ at time moment t
  E   ▷ Average prediction errors
  D   ▷ Average absolute prediction errors
  minFH   ▷ Relabel prediction horizon

1: **while** $(FH[Strt] < m)$ **do**
2:     Strt++;
3:     PY[Strt][0]=Y[Strt];
4:     PY[Strt][1]=Y[Strt+1];
5:     **for all** $(t = Strt + 2, \ldots m)$ **do**
6:         py=0;
7:         **for all** $j = 0, 1, \ldots n$ **do**
8:             A1=G[j](PY[Strt][t-1],PY[Strt][t-2]);
9:             R=a[j]*A1;
10:            py+=R;
11:        **end for**
12:        PY[Strt][t]=py;
13:        **if** $(|PY[Strt][t] - Y[(Strt) + t]| > SZ)$ **then**
14:            **Break;**
15:        **end if**
16:    **end for**
17:    FH[Strt]=t;
18: **end while**
19: LastStrt=t;
20: minFH=FH[Strt];
21: int minFHp=minFH;
22: **for all** $t = 3, \ldots Strt$ **do**
23:    **if** (minFH > FH[t]) **then**
24:        minFHp=FH[t];
25:    **end if**
26: **end for**
27: minFH=(minFHp<minFH)? minFHp : minFH;
28: E=D=0;   ▷ minFHp is the reasonable horizon
29: **for all** $t = 3, \ldots minFH$ **do**
30:    D+=fabs(Y[t+Strt]-PY[Strt][t]);
31:    E+=(Y[t+Strt]-PY[Strt][t]);
32: **end for**
33: D/=minFH; E/=minFH;
   return (D; E; minFH)

---

series with 888 points for the daily infected cases of Covid-19 in Chelyabinsk region. The graph of this process is shown in Fig. 3. We see that this process in

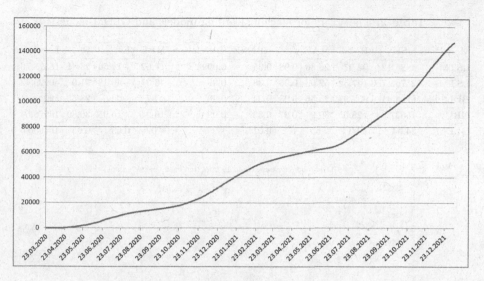

Fig. 2. The observed time series

Table 1. The identification results

Timeframe	150	300	500
a_1	1.982416e00	2.056573e00	1.999967e00
a_2	−9.824160e−01	−1.056573e00	−9.999669e−01
a_3	7.459095e01	−4.031902e01	−2.774247e00
a_4	7.606037e01	−4.092965e01	−2.805568e00
a_5	−1.506450e02	8.124591e01	5.579729e00
Loss function	2.811987e−01	5.663879e−02	3.828931e−02
Reasonable forecasting horizon	149	299	499
E		−3.640266e−03	9.866220e−01	9.919839e−01
D	9.350328e−01	9.866220e−01	9.919839e−01

neither monotonous nor oscillate, it has both small as very high peaks in different periods. We see that the model exactly reproduced this process by the obtained coefficients and allowed to get a forecast for 14 days. We used this time series to run machine learning algorithms and calculated the errors and the value of loss function for them (see Table 2). As we can see from this table the proposed model allows get better results, and lower errors, and the $R^2 = 1.0$ that means that this model is very effective for capturing the pattern of the data. The other examples and full source code is published in [14].

The main aim to identify the equations is enabling to use the model values of endogenous variables for forecasting possible values of corresponding endogenous variables in the future. Algorithm 4 allows to estimate the reasonable forecasting horizon, and for the considered example it is one day less then the length of initial

Table 2. The errors for machine learning models and GLDM-model

Model	MSE	RMSE	MAE	R^2	RRMSE	Correlation	MBE	E	D	Loss Func
LSTM	8886.12	94.27	26.86	0.98	0.25	0.99	17.62	−17.56	26.81	15571.76
LSTMs	21761.75	147.52	52.03	0.96	0.38	0.99	46.3	−46.18	51.9	40929.42
BDLSTM	8703.41	93.29	28.56	0.98	0.24	0.99	0.67	−0.66	28.55	590.96
GRU	15637.84	125.05	30.9	0.97	0.33	0.99	16.94	−16.95	30.89	14978.54
GLDM	34.04	5.83	0.65	1	0.01	1	−0.65	0.65	0.65	165950.5

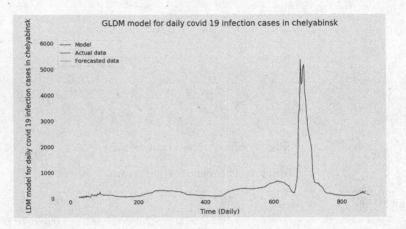

Fig. 3. The results of running GLDM model for daily Covid-19 infection cases in Chelyabinsk region (888 days)

data. It is not so for some other time series, especially for time series of oscillatory process with minimums and maximums on the observed period. From Table 1 it is easy to see that even for very large forecasting horizon the average errors have low values and they do not depend on the length of initial vector.

The considered computational experiment is based on standard numeric data types and does not use parallelization. Perhaps, either increasing the order of difference equation, or increasing of vector length (making the algorithm running for big data arrays) may cause using special data types for increasing the accuracy or using parallel technologies for speeding up the algorithm execution.

6   Conclusion

Speaking about the quality of the model itself we can mention that it works not worse than neural network models or classical statistical models. It has one significant advantage in comparison with these models that is in the opportunity to interpret the model coefficients in term of the research problem. The method considered in the article is another alternative to the construction of digital twins of the production process. Unlike neural networks, this approach makes it possible to explicitly obtain high-quality quasi-linear difference equations (adequately describing the considered process). Directions of further researches are the use of the above algorithms for forecasting the multidimensional time series, and many other points.

References

1. Ponce, M.: Covid 19 analitic: an R package to obtain, analyze and visualize data from the corona virus disease pandemic (2020). arXiv:2009.01091v1
2. Makarovskikh, T., Abotaleb, M.: Comparison between two systems for forecasting Covid-19 infected cases. IFIP Adv. Inf. Commun. Technol. **616**, 107–114 (2021). https://doi.org/10.1007/978-3-030-86582-5_10
3. Neto, A., Ferreira, T., Batista, M., Firmino, P.: Studying the performance of cognitive models in time series forecasting. Revista de Informatica Teorica e Aplicada 27(1), 83–91 (2020). https://doi.org/10.22456/2175-2745.96181
4. Pan, J., Wang, H., Qiwei, Y.: Weighted least absolute deviations estimation for ARMA models with infinite variance. Economet. Theor. **23**(3), 852–879 (2007)
5. Panchal, R., Kumar, B.: Forecasting industrial electric power consumption using regression based predictive model. Recent Trends Commun. Electron. (2021). https://doi.org/10.1201/9781003193838-26
6. Panyukov, A.V., Mezaal, Y.A.: Stable estimation of autoregressive model parameters with exogenous variables on the basis of the generalized least absolute deviation method. IFAC-PapersOnLine **51**(11), 1666–1669 (2018). https://doi.org/10.1016/j.ifacol.2018.08.217. Open access
7. Panyukov, A.V., Mezaal, Y.A.: Improving of the identification algorithm for a quasilinear recurrence equation. In: Olenev, N., Evtushenko, Y., Khachay, M., Malkova, V. (eds.) OPTIMA 2020. CCIS, vol. 1340, pp. 15–26. Springer, Cham (2020). https://doi.org/10.1007/978-3-030-65739-0_2
8. Panyukov, A.: Scalability of algorithms for arithmetic operations in radix notation. Reliable Comput. **19**, 417–434 (2015). http://interval.louisiana.edu/reliable-computing-journal/volume-19/reliable-computing-19-pp-417-434.pdf
9. Panyukov, A., Mezaal, Y.: Stable identification of linear autoregressive model with exogenous variables on the basis of the generalized least absolute deviation method. Bull. South Ural State Univ. Ser. Math. Model. Program. Comput. Softw. 11(1), 35–43 (2018). https://doi.org/10.14529/mmp180104
10. Panyukov, A., Tyrsin, A.: Stable parametric identification of vibratory diagnostics objects. J. Vibroeng. **10**(2), 142–146 (2008). https://www.extrica.com/article/10181
11. Sirotin, D.: Neural network approach to forecasting the cost of ferroalloy products. Izvestiya. Ferrous Metall. 63(1), 78–83 (2020). https://doi.org/10.17073/0368-0797-2020-1-78-83
12. Tyrsin, A.N.: Robust construction of regresson models based on the least absolute deviations method. J. Math. Sci. **139**(3), 6634–6642 (2006). https://doi.org/10.1007/s10958-006-0380-7
13. Yakubova, D.: Econometric models of development and forecasting of black metallurgy of Uzbekistan. Asian J. Multidimensional Res. (AJMR) **8**, 310–314 (2019). https://doi.org/10.5958/2278-4853.2019.00205.2
14. Abotaleb M.: GLDM-model (2022). https://github.com/abotalebmostafa11/GLDM-model

Author Index

Printed in the United States
by Baker & Taylor Publisher Services